Review of the Marine Recreational Information Program

Committee on the Review of the
Marine Recreational Information Program

Ocean Studies Board

Division on Earth and Life Studies

A Consensus Study Report of

The National Academies of
SCIENCES · ENGINEERING · MEDICINE

THE NATIONAL ACADEMIES PRESS
Washington, DC
www.nap.edu

THE NATIONAL ACADEMIES PRESS 500 Fifth Street, NW Washington, DC 20001

This study was supported by the National Oceanic and Atmospheric Administration under Award Number WC133R-11-CQ-0048, TO #10. Any opinions, findings, conclusions, or recommendations expressed in this publication do not necessarily reflect the views of any organization or agency that provided support for the project.

International Standard Book Number-13: 978-0-309-45374-5
International Standard Book Number-10: 0-309-45374-7
Digital Object Identifier: https://doi.org/10.17226/24640

Additional copies of this publication are available for sale from the National Academies Press, 500 Fifth Street, NW, Keck 360, Washington, DC 20001; (800) 624-6242 or (202) 334-3313; http://www.nap.edu/.

Printed in the United States of America

Suggested citation: The National Academies of Sciences, Engineering, and Medicine. 2017. *Review of the Marine Recreational Information Program*. Washington, DC: The National Academies Press. doi: https://doi.org/10.17226/24640.

The National Academies of

SCIENCES • ENGINEERING • MEDICINE

The **National Academy of Sciences** was established in 1863 by an Act of Congress, signed by President Lincoln, as a private, nongovernmental institution to advise the nation on issues related to science and technology. Members are elected by their peers for outstanding contributions to research. Dr. Marcia McNutt is president.

The **National Academy of Engineering** was established in 1964 under the charter of the National Academy of Sciences to bring the practices of engineering to advising the nation. Members are elected by their peers for extraordinary contributions to engineering. Dr. C. D. Mote, Jr., is president.

The **National Academy of Medicine** (formerly the Institute of Medicine) was established in 1970 under the charter of the National Academy of Sciences to advise the nation on medical and health issues. Members are elected by their peers for distinguished contributions to medicine and health. Dr. Victor J. Dzau is president.

The three Academies work together as the **National Academies of Sciences, Engineering, and Medicine** to provide independent, objective analysis and advice to the nation and conduct other activities to solve complex problems and inform public policy decisions. The National Academies also encourage education and research, recognize outstanding contributions to knowledge, and increase public understanding in matters of science, engineering, and medicine.

Learn more about the National Academies of Sciences, Engineering, and Medicine at **www.nationalacademies.org.**

The National Academies of
SCIENCES · ENGINEERING · MEDICINE

Consensus Study Reports published by the National Academies of Sciences, Engineering, and Medicine document the evidence-based consensus on the study's statement of task by an authoring committee of experts. Reports typically include findings, conclusions, and recommendations based on information gathered by the committee and the committee's deliberations. Each report has been subjected to a rigorous and independent peer-review process and it represents the position of the National Academies on the statement of task.

Proceedings published by the National Academies of Sciences, Engineering, and Medicine chronicle the presentations and discussions at a workshop, symposium, or other event convened by the National Academies. The statements and opinions contained in proceedings are those of the participants and are not endorsed by other participants, the planning committee, or the National Academies.

For information about other products and activities of the National Academies, please visit www.nationalacademies.org/about/whatwedo.

COMMITTEE ON THE REVIEW OF THE MARINE RECREATIONAL INFORMATION PROGRAM

LUIZ BARBIERI, *Co-Chair,* Florida Fish and Wildlife Conservation Commission, St. Petersburg
CYNTHIA M. JONES, *Co-Chair,* Old Dominion University, Norfolk, Virginia
JILL A. DEVER, RTI International, Washington, D.C.
DAVID HAZIZA, Université de Montréal, Montreal, Quebec, Canada
JEFFREY C. JOHNSON, University of Florida, Gainesville
BRUCE M. LEAMAN, International Pacific Halibut Commission, Seattle, Washington
THOMAS J. MILLER, University of Maryland Center for Environmental Science, Solomons
SEAN P. POWERS, University of South Alabama, Mobile
STEVE WILLIAMS, Pacific States Marine Fisheries Commission, Portland, Oregon

Staff

STACEE KARRAS, Program Officer, Ocean Studies Board
DAVID POLICANSKY, Scholar, Board on Environmental Studies and Toxicology
MICHAEL COHEN, Senior Program Officer, Committee on National Statistics
PAYTON KULINA, Senior Program Assistant, Ocean Studies Board
ALLIE PHILLIPS, Program Assistant, Ocean Studies Board

OCEAN STUDIES BOARD

Preface

In 2004, the National Oceanic and Atmospheric Administration's National Marine Fisheries Service (NMFS) requested that the National Research Council (NRC; now referred to as the National Academies of Sciences, Engineering, and Medicine, or "the National Academies"), review data collection for marine recreational fisheries in the United States, and specifically the Marine Recreational Fisheries Statistics Survey (MRFSS). The NRC formed a committee comprising 10 experts in fishery science and statistics, which released its report, *Review of Recreational Fisheries Survey Methods*, in 2006. Together, the recommendations of the 2006 report called for a considerable redesign of the entire survey program to update survey methods to reduce bias, increase efficiencies, and allow for greater stakeholder trust and better relations with the recreational angling community.

Since 2007, in response to the NRC report, NMFS has worked to improve the survey program, primarily by transitioning from the MRFSS to the redesigned Marine Recreational Information Program (MRIP). A decade after the release of the 2006 report, NMFS requested the National Academies conduct a second study to evaluate how well and to what extent NMFS has addressed the NRC's recommendations. The current report is a result of this latest effort.

The need for this evaluation is clear. Provisions in the 2006 reauthorization of the Magnuson-Stevens Fisheries Conservation and Management Act greatly increased the demand for high-quality and timely data that can be used for assessment and management of marine fish stocks. However, because of the shortcomings of the MRFSS, NMFS faced a lack of confidence in providing the quality data needed for managing recreational fisheries. Having an independent and objective review of the progress made since implementation of the

MRIP should address many of the previous concerns and help reassure anglers and stakeholders.

Producing this report was a difficult challenge because of the complexity and multidisciplinary nature of the issues involved. Throughout the study, NMFS was always responsive to the committee's numerous questions and requests for information. In particular, we thank the MRIP staff for their patience and openness in addressing questions about the program, and Dr. Ned Cyr, director of the Office of Science and Technology, for setting the stage for this review.

The committee is also grateful to the many individuals who played a role in the completion of this study. The committee met four times and would like to extend its gratitude to all the individuals from regional councils, state fisheries agencies, recreational and commercial fisheries organizations, environmental conservation organizations, and others who appeared before the full committee or otherwise provided background information and discussed pertinent issues.

Finally, the committee sincerely thanks the National Academies' staff for their valuable support and extra efforts to facilitate the rapid completion of the report without compromising quality: Stacee Karras (Study Director), David Policansky (Scholar), Michael Cohen (Senior Program Officer), Payton Kulina (Senior Program Assistant), and Allie Phillips (Program Assistant).

Cynthia Jones and Luiz Barbieri, *Committee Co-Chairs*

Acknowledgments

The committee would especially like to thank the National Marine Fisheries Service staff and contractors for their invaluable assistance in providing background documents requested by the committee and for their participation in meetings and on calls. In particular, the committee thanks Ned Cyr, David Van Voorhees, Gordon Colvin, Tom Sminkey, John Foster, Rob Andrews, Leah Sharpe, Lauren Dolinger Few, and David Bard.

This report was also greatly enhanced by discussions with participants at the committee's meetings as part of this study. The committee would like to especially acknowledge the efforts of those who gave presentations at these meetings: Rob Andrews (NOAA), Michael Armstrong (Massachusetts Division of Marine Fisheries), David Bard (NOAA Contractor), Tom Baum (New Jersey Department of Fish and Wildlife), Bob Beal (Atlantic States Marine Fisheries Commission), Harry Blanchet (Louisiana Department of Wildlife and Fisheries), David Blazer (Maryland Department of Natural Resources), John Boreman (Mid-Atlantic Fishery Management Council, Scientific and Statistical Committee), Zack Bowen (Charter Boat Captain), Richen Brame (Coastal Conservation Association), Gregg Bray (Gulf States Marine Fisheries Commission), Jay Breidt (Colorado State University), Kenneth Brennan (NOAA), Mike Brown (California Department of Fish and Wildlife), John Carmichael (SAFMC), Brittany Chudzik (Mississippi Department of Marine Resources), Gordon Colvin (NOAA Contractor), Roy Crabtree (Southeast Regional Office), Ned Cyr (NOAA), E.J. Dick (Southwest Fisheries Science Center, NOAA), Lauren Dolinger Few (NOAA), Michelle Duval (South Atlantic Fishery Management Council), Daniel Erickson (Oregon Department of Fish and Wildlife), Mark Fisher (Texas Parks and Wildlife), Brad Floyd (South Carolina Department of Natural Resources), John Foster (NOAA),

John Froeschke (Gulf of Mexico Fishery Management Council), Chad Hanson (The Pew Charitable Trusts), Michael Kelly (CLS America), Kathy Knowlton (Georgia Coastal Resources Division), Mike Leonard (American Sportfishing Association), Chris Macaluso (Theodore Roosevelt Conservation Partnership), Richard Merrick (NOAA), Doug Mumford (North Carolina Division of Marine Fisheries), Corey Niles (Washington Department of Fish and Wildlife), Jean Opsomer (Colorado State University), Todd Phillips (The Ocean Conservancy), Clay Porch (Southeast Fisheries Science Center), Beverly Sauls (Florida Fish and Wildlife Conservation Commission), Leah Sharpe (NOAA Contractor), Tom Sminkey (NOAA), Lynne Stokes (Southern Methodist University), David Van Voorhees (NOAA), Geoff White (Access Point Angler Intercept Survey), and Dan Wolford (Coastside Fishing Club).

This Consensus Study Report was reviewed in draft form by individuals chosen for their diverse perspectives and technical expertise. The purpose of this independent review is to provide candid and critical comments that will assist the National Academies of Sciences, Engineering, and Medicine in making each published report as sound as possible and to ensure that it meets the institutional standards for quality, objectivity, evidence, and responsiveness to the study charge. The review comments and draft manuscript remain confidential to protect the integrity of the deliberative process.

We thank the following individuals for their review of this report:

Mike Brick, Westat, Inc., Rockville, MD
Michele Culver, Washington Department of Fish and Wildlife, Montesano
Bonnie McCay, Rutgers, The State University of New Jersey, Stockton
Steven Murawski, University of South Florida, St. Petersburg
Louis-Paul Rivest, Université Laval, Quebec, Canada
Kenneth Rose, Louisiana State University, Baton Rouge
Carl Schwarz, Simon Fraser University, Burnaby, BC, Canada
Patrick Sullivan, Cornell University, Ithaca, NY
Geoff White, Atlantic Coastal Cooperative Statistics Program, Arlington, VA

Although the reviewers listed above provided many constructive comments and suggestions, they were not asked to endorse the conclusions or recommendations of this report nor did they see the final draft before its release. The review of this report was overseen by **John Dowling**, Harvard University, and **Andrew Solow**, Woods Hole Oceanographic Institution. They were responsible for making certain that an independent examination of this report was carried out in accordance with the standards of the National Academies and that all review comments were carefully considered. Responsibility for the final content rests entirely with the authoring committee and the National Academies.

Contents

Appendixes

Summary

The National Marine Fisheries Service (NMFS) of the National Oceanic and Atmospheric Administration is responsible for collecting information on marine recreational angling. It does so principally through the Marine Recreational Information Program (MRIP), a survey program that consists of an in-person survey at fishing access sites and a mail survey, in addition to other complementary or alternative surveys. Data collected from anglers through the MRIP supply fisheries managers with essential information for assessing fish stocks. In 2006, the National Research Council (NRC; now referred to as the National Academies of Sciences, Engineering, and Medicine, or "the National Academies") provided an evaluation of the MRIP's predecessor, the Marine Recreational Fisheries Statistics Survey (MRFSS). That review, *Review of Recreational Fisheries Survey Methods*, presented conclusions and recommendations in six categories: sampling issues, statistical estimation issues, human dimensions, program management and support, communication and outreach, and general recommendations.

After spending nearly a decade addressing the recommendations, NMFS requested another evaluation of its modified survey program (MRIP). This report, the result of that evaluation, serves as a 10-year progress report (Box S.1). The committee met on four occasions, in Washington, DC; Charleston, South Carolina; New Orleans, Louisiana; and Irvine, California. At each meeting, the committee heard from state and federal employees as well as regional stakeholders. The committee also received written input from stakeholders during the study process. The resulting report recognizes the impressive progress that NMFS has made, including major improvements in the statistical soundness of its survey designs. It also highlights some remaining challenges, and provides recommendations for addressing them. This report principally focuses on the Fishing Effort

BOX S.1
Statement of Task

An ad hoc NRC committee will assess progress in updating marine recreational fisheries data collection through the Marine Recreational Information Program (MRIP) referencing the recommendations in the 2006 NRC report *Review of Recreational Fisheries Survey Methods*. Based on this assessment, the committee will identify potential areas for improvements or changes of direction that would substantially increase data quality for fisheries management, taking into consideration potential loss of information from disruption of the time series. The committee's report will

1. Describe the approach and effectiveness of steps taken by NMFS to improve the quality and accuracy of marine recreational fisheries catch, effort, and participation statistics (in response to NRC, 2006), including, but not limited to,
    a. Establishing registries of anglers and for-hire vessels and using the registries appropriately as sample frames for recreational catch and effort surveys;
    b. Improving the effectiveness and appropriateness of sampling and estimation procedures, applicability to various kinds of management decisions, and usefulness for social and economic analyses; and
    c. Providing for ongoing technical evaluation and modification, as needed to meet emerging management needs and changes in communication technologies (e.g., smartphone apps, Internet-based social networking).
2. Assess the strength of the scientific process, including the engagement of external scientific and technical expertise, used by NMFS in developing, testing, reviewing, and certifying new sampling and estimation procedures.
3. Evaluate the communication of information on survey method development, survey method descriptions, and survey results to stakeholders and application of stakeholder input in the design and implementation of new sampling and estimation procedures. Stakeholders include at least three distinct subgroups (with some overlap among them):
    a. Data-collection partners, such as the Atlantic Coast Cooperative Statistics Program and the Fishery Information Networks;
    b. Data customers (parties that use NMFS data for stock assessments, management actions, social and economic studies); and
    c. Entities affected by the estimates (anglers and recreational fishing businesses, commercial fisheries, nonconsumptive users, etc.).
4. Determine if the degree of coordination among federal, state, and territorial survey programs is sufficient to provide a clear, national perspective on marine recreational fisheries.
5. Evaluate plans for maintaining continuity of data series to minimize disruption of management programs and stock assessments. This will include evaluation of the strategy for moving from the phone-based survey to a mail- and web-based survey as a means to estimate fishing effort.

Survey (FES) and the Access Point Angler Intercept Survey (APAIS) that form the backbone of the MRIP, but recognizes the role that other survey programs play in the MRIP and discusses them as they relate to coordination, certification, continuity, and other important aspects of the MRIP.

FISHERIES SURVEYS AND MANAGEMENT

Recreational fishing is a favorite pastime in the United States. Recreational anglers throughout the nation fish from beaches and piers, as well from private, rental, or charter boats. Although recreational anglers each may take only a small number of fish, collectively, they can have a significant impact on the overall abundance of a stock. In some fisheries, the recreational catch exceeds the commercial catch.

Several attributes of recreational fisheries make them difficult to assess. Recreational fisheries include a large number of participants using many fishing modes at or from many diverse access points. Some recreational anglers travel great distances to fish, while others fish from their private property. Some anglers often fish, while others seldom fish. These characteristics make recreational anglers difficult to characterize and monitor. To further complicate matters, recreational anglers release some of their catch, and because discard mortality is difficult to assess, the overall impact on fish stocks also is difficult to assess.

MARINE RECREATIONAL FISHERIES STATISTICS SURVEY

In 1979, NMFS established the MRFSS as a national program for gathering standardized and comparable data on marine recreational fisheries in the United States. The MRFSS primarily consisted of two independent but complementary surveys, a coastal household telephone survey using random-digit dialing to sample potential anglers, and an in-person intercept survey that sampled anglers where they completed their fishing trips.

Data collected from telephone surveys were used to produce estimates of effort (i.e., number of angler trips taken), and data collected from intercept surveys were used to establish estimates of catch per unit effort (CPUE). Estimates of total catch are derived using the product of effort and CPUE. Catch estimates are crucial for stock assessment and management (i.e., to avoid overfishing or to rebuild overfished stocks).

In addition to the telephone and intercept surveys, the MRFSS program allowed for alternative or supplemental region-, state-, species-, or sector-specific surveys. Alaska has never been part of the MRFSS program, and Texas has not been since 1985; both conduct their own surveys.

Since the MRFSS was established, the context for conducting marine recreational fishing surveys has changed. Demand for active management on narrower temporal and spatial scales has increased, and the mix of recreational and commercial fishing has changed for some species and regions. By the early 2000s, many anglers, managers, and fishery scientists were concerned that the use of data produced by the MRFSS in management exceeded the original design and purposes of the MRFSS. Specifically, they were concerned that the precision, robustness, and timeliness of data were misaligned with management needs. Social and technological changes were also impacting the surveys' effectiveness and efficiency.

The conclusions and recommendations from the 2006 NRC report called for a considerable redesign of the survey program to reduce bias, increase efficiencies, and allow for greater stakeholder relations. However, the 2006 report also acknowledged the considerable complexity and challenges associated with such changes and supported making additional resources available for this purpose. In 2007, Congress called for implementation of the recommendations in the report to the extent feasible.

MARINE RECREATIONAL INFORMATION PROGRAM

Since 2007, NMFS has worked to improve the survey program by transitioning from the MRFSS program to the MRIP. Like the MRFSS program, the MRIP is composed predominantly of an intercept survey (APAIS) to gauge CPUE and a separate offsite effort survey (FES) to determine effort. However, both surveys have undergone significant changes in terms of methodologies and statistical analyses. For example, the offsite FES has been transitioning from the telephone survey to a mail-based survey that employs address-based sampling.

The MRIP also funds a variety of region-, state-, species-, and sector-specific surveys that either supplement or serve as alternatives to the APAIS and FES (see Figure S.1). NMFS has had to consider how to continue to provide flexibility for these other surveys, which are tailored for specific circumstances, while retaining sufficient data consistency to maintain a national perspective.

Fishing Effort Survey

Sample Design and Data Collection

Fishing effort has historically been estimated with data collected from the telephone survey. The 2006 report cited a growing number of issues affecting the bias and precision of estimates. These included potentially low data quality because of undercoverage bias from increasingly fewer households having landline phones, in addition to already low response rates, which were projected to further decrease over time. That report suggested a national angler registry as a possible solution, because it could serve as a list from which to sample (also referred to as a sample frame).

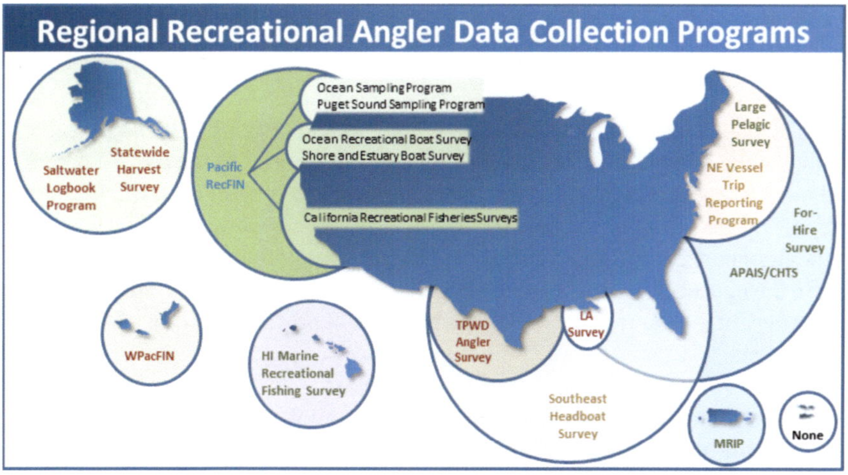

FIGURE S.1 A visualization of where various recreational fisheries surveys are implemented within the United States. Most are at least in part supported by the MRIP. The Texas Parks and Wildlife Department survey and both surveys conducted in Alaska, however, are not supported by the MRIP funds. Represented in the individual circles (from left to right) are Alaska, Guam and Samoa, Hawaii, Puerto Rico, and the U.S. Virgin Islands. SOURCE: NMFS, 2014a.

The 2007 reauthorization of the Magnuson-Stevens Fishery Conservation and Management Act required NMFS to create the National Saltwater Angler Registry (NSAR). The statute and regulations provide an exemption for states with saltwater license registries. State license frames could serve to meet federal requirements. Pilot studies conducted by NMFS indicated that the NSAR is not an ideal sample frame, however, because most states have exemptions in their license requirements, and therefore, coverage is not uniform. Instead, NMFS developed an innovative mail survey that uses address-based sampling enhanced by the NSAR to improve effectiveness and appropriateness of fishing effort estimation. A pilot test with this frame resulted in impressive improvements over the telephone survey used in the MRFSS, and the committee commends NMFS's innovative use of the registry. This important shift from a phone survey to a mail survey also addresses societal trends such as the increasing reliance on cell phones and declining use of landlines. The enhanced sampling frame enabled a direct link to coastal households through geolocation information. Additionally, this new approach provides another level of stratification for sampling associated with license status (Yes versus No/Unknown). The methodologies associated with the current FES, including the address-based sampling mail survey design, are major improvements from the original Coastal Household Telephone Survey.

Survey Material

The mail survey includes a relatively short questionnaire, a cover letter, frequently asked questions, a prepaid return envelope, and a small cash incentive. The questionnaire contains 10 household-wide questions regarding whether anyone in the household has been fresh- or saltwater fishing in the past 12 months and other household details. Six questions regarding demographics, whether they saltwater fished from shore or boat, and the number of days fished by location in the designated preceding 2-month period are asked of at most five individuals within the household. Adding a question on the use of public versus private access points may help discern whether fishing behavior at private sites varies from those at public sites.

Recommendation: NMFS should conduct pilot studies to determine the optimal method for collecting accurate information on total catch differences between public- and private-access sites. For example, NMFS could add a question to the Fishing Effort Survey questionnaire about angler use of private sites or public access sites. Geographic maps used to identify public access sites within the state could be used to distinguish public access from private sites.

The 2-month recall period was set for consistency with the periods covered by the intercept survey. Several factors, however, determine the anglers' ability to recall the number of fishing trips and the number and types of fish caught.

Recommendation: NMFS should consider evaluating a prospective data-collection methodology, such as asking people in advance to document fishing trips planned over the next 2 months, to reduce concerns about angler recall.

Data Quality

Changes in fishing effort trends may suggest problems with nonresponse bias or quality of the responses, or could indicate actual change over time. However, without additional data, explanations can only be speculative.

Recommendation: As recommended in the 2006 report, NMFS is encouraged to continue research on survey panels, where a portion of the sampled households is retained for one or more interviews, for the Fishing Effort Survey alone or for an effort-catch combined study. The purpose of the survey panel would be to assess trends and any anomalies in those trends, to assess any improvements in data-collection efficiency through increased participation, and possibly to lower measurement

error associated with, for example, trip recall with a more engaged sample of anglers.

Web questionnaires and cell phone applications may be a viable option to increase production of fishing effort estimates with data that are evaluated in real time.

Recommendation: As recommended in the 2006 report, electronic data collection should be further evaluated as an option for the Fishing Effort Survey, including smartphone apps, electronic diaries for prospective data collection, and a web option for all or just panel members.

Missing responses also lower the quality of the fishing effort estimates. Missing items from an otherwise complete questionnaire can be addressed with imputation, where the missing value is replaced with a valid response using a defined model. In addition, as an enhancement to the standard FES design, NMFS could evaluate a nonresponse follow-up, where a random subsample of nonresponding households is recontacted using, for example, an additional incentive to improve response and to evaluate nonresponse bias.

Recommendation: NMFS should consider conducting targeted annual nonresponse studies as a standard component of the MRIP. The purpose of these studies would be to continually monitor correlates of nonresponse and nonresponse bias to control its damaging effects on data quality.

Weighting and Estimation

The FES weighting methodology includes key components to reduce sampling and nonsampling errors in the estimates. NMFS could additionally evaluate a separate unknown-eligibility weight adjustment for nonresponding households. Furthermore, collaboration with other federal agencies to obtain estimates of anglers to enhance the poststratification methods can improve data quality.

Precision for estimated fishing effort in the FES is calculated with methods that only account for a portion of the adjustment applied to the analysis weights. Thus, the sampling variance for the estimates may be too small.

Recommendation: Other variance estimation methods should be evaluated for fishing effort estimates to account for weight adjustments, especially those associated with nonresponse. These include replication methods and the so-called reverse approach.

Access Point Angler Intercept Survey

The second component of the MRIP is the APAIS. Although the MRFSS also relied on an intercept survey, it lacked a proper statistical foundation. The APAIS is also an onsite survey, but it accounts for the probabilistic aspects of the intercepts such that the onsite surveys now have a solid statistical basis.

Data Collection

The two main data-collection tasks of the APAIS are counts of completed angler fishing trips and angler-intercept interviews. The angler interviews are obtained by intercepting marine recreational anglers at shore or boat access points after they have been fishing. Interviewers obtain information about the completed trip, including fishing locations, the species and number of fish caught, the gear used, and the length of the trip. Interviewers have the opportunity to observe, weigh, and measure the length of the catch, also confirming species identification. Sampling in the party (or head) boat mode may include having observers on the boats. The interviewers obtain some demographic data about the anglers and record the number of fish that were released and not landed. Interviewers are instructed to count all anglers completing their trips, even those that are not interviewed.

The APAIS sampling frame and site registry are major improvements from the MRFSS. Each interviewer's assignment now consists of a fixed time interval at a particular site, with the frequency of sampling that time and place being dependent on the historical number of trips, such that the busiest sites and times are sampled with the greatest frequency. Interviewers attempt to obtain the largest possible number of completed interviews for a given assignment. In a major improvement over the MRFSS, the MRIP interviewers are strictly scheduled, all time periods of the day are eligible to be sampled including nights, and there are no caps on the number of interviews that samplers should take. Interviewers are no longer allowed discretion of which sites to sample. Additionally, field staff visit sites and update the site registry periodically to ensure that the registry is current and covers all public sites.

Some data are missed, because some anglers refuse to be interviewed or refuse to answer particular questions, or because of language barriers. Anglers might also be missed if there are too many at the access point at the same time for all to be interviewed. Collecting as much information as possible about these nonresponding anglers may help explain refusals and address concerns that such parties have a different CPUE than the responding anglers. As was noted in the 2006 report, because private access sites generally cannot be sampled, the use of CPUE from public-access sites for the calculation of total catch rests on the strong assumption that private-access CPUE and target species do not differ from public access. As noted above, NMFS could add a question to the FES to ascertain public or private access.

Survey Material

Each interview is conducted with a scripted questionnaire that records catch, release, and trip information. The responses are mostly recorded on paper then subsequently coded and entered into a database to be quality controlled for out-of-range answers. The committee sees value in moving to electronic recording of these data, which will improve acquisition time for managers and permit immediate quality control of input. Some anglers are also eager to input their data to the MRIP regardless of whether they are intercepted. Although this is admirable, such nonprobabilistic sampling can be highly unrepresentative of the general angling public and hence statistically problematic. The MRIP and its consultants have investigated approaches that might help resolve the considerable statistical difficulties in using such data, and the committee encourages continuation of this endeavor into the future.

Because onsite interviews are conducted in person, there is opportunity to clarify the questions asked of anglers. Moreover, the interviewer observes the catch and the number of trips so there is no problem with recall bias as there is in the FES. However, this also means that the interviewer is the public face of the MRIP. With oversight by NMFS, interviewer training is largely done by state partners, and along the Atlantic coast, by the Atlantic Coastal Cooperative Statistics Program. State agency personnel now conduct interviews, thus enhancing confidence in this component of the survey. Still, good interview training is critical.

Data Quality

The statistical soundness of the intercept survey has been considerably strengthened since the 2006 review due to the previously discussed improvements to data-collection methods. The APAIS also provides valuable information on the number of anglers that are intercepted who reside in noncoastal households. These data are used to scale up the effort estimates from the FES. Still, challenges remain, including difficulty in estimating the number, species, and fates of fish released rather than landed, and the difficulty of dealing with private-access sites, which cannot be sampled and therefore must be estimated. Generally, the statistical validity of the survey can be further strengthened through additional analyses, obtaining some small amount of additional information in the interviews, and improving methods for estimation and validation of the numbers, species, and fates of fish discarded by anglers.

Weighting and Estimation

With the new APAIS design, the inclusion probabilities corresponding to angler trips can be easily computed. The weights used in the estimation procedures are obtained as the inverse of the inclusion probabilities. The resulting point and variance estimators are, to a good approximation, design unbiased.

FRAMEWORK FOR CONTINUED SCIENTIFIC EVALUATION, REVIEW, AND CERTIFICATION

In addition to providing specific recommendations for improving the effort and intercept surveys, the 2006 report identified the need for an improved framework for continual scientific evaluation, review, and certification of the methods, protocols, and procedures used for data collection. NMFS has made substantial progress toward such a framework. As the MRIP's focus evolves from developing and testing survey improvements to increasingly putting new methods to practice in the field, the timeliness of the survey review and certification process could benefit from additional attention.

Capacity and Scientific Evaluation

The 2006 report recommended that a survey office devoted to the management and implementation of marine recreational surveys be developed. Since 2006, the number of MRIP staff has increased from 6 to 12 full-time staff. The program has also invested in formal training of existing staff, including providing opportunities to earn advanced degrees or take courses in topics such as survey methodology. The MRIP's staff expansion appears to have greatly increased its ability to expand technical support and achieve better regional coordination.

The MRIP has benefited greatly from the independent research group of statisticians and survey methodologists who not only assess the general adequacy of the MRIP but also provide technical advice to regional and state programs. If NMFS is able to expand the existing capacity in this pool of consultants both in number and in expertise (e.g., experts in cognitive issues, including angler recall), duplication of effort would be reduced and the provision of technical and scientific support would be facilitated. In addition, the MRIP certification process would be streamlined. Any such group would further benefit from being periodically refreshed to include new researchers with a variety of interests and expertise.

The MRIP has either organized or been involved in the organization of several workshops or symposia, which have been attended by highly trained statisticians and fishery scientists. These meetings have facilitated review and discussion of MRIP issues by a broad range of experts, promoting an exchange of ideas, and giving MRIP technical staff, as well as regional and state partners, an opportunity to explore a variety of recreational fisheries issues under different scenarios. The committee commends the MRIP for this outreach.

Pilot Projects

In 2008, the MRIP established a pilot studies program for developing, testing, reviewing, and eventually certifying new sampling and estimation procedures to be applied under the MRIP umbrella, mostly in collaboration with state and regional partners. The MRIP pilot studies program is implemented in three concurrent phases: (1) evaluation of current methods, (2) innovation to identify and test new methods, and (3) implementation of proven methodologies. The MRIP Operations Team solicits and reviews research proposals and provides recommendations for funding. The program constitutes an appropriate and effective mechanism for providing highly specialized technical and scientific support toward the development, review, and certification of surveys.

Use of New Technology

Traditionally, recreational fishing survey responses have been recorded using paper-and-pencil survey forms. However, recently there has been a great deal of interest in the recreational fishing community in identifying scientifically sound, statistically robust methods for using electronic reporting (e.g., using smartphones and tablets). These new technologies could potentially improve the timeliness and accuracy of data and reduce costs and paperwork burdens.

Electronic data collection could be integrated into the MRIP in four separate and distinct ways:

1. Use of electronic logbooks by the for-hire sector,
2. Enabling interviewers to capture and submit data electronically,
3. Allowing anglers to self-report data electronically, and
4. Using electronic monitoring to validate self-reported data.

Evaluation and testing of new technologies for MRIP fisheries data collection is being accomplished through several MRIP-funded pilot studies, often structured according to Regional MRIP Implementation Plans. Despite these efforts, for portions of the private angler and for-hire sectors that implementation of electronic reporting is not occurring fast enough.

Recommendation: The MRIP should develop a strategy to better articulate the complexities, costs, and timelines associated with implementing new and emerging technologies in recreational fisheries data collection and monitoring. This communication strategy should not only focus on regional partners but also address questions and concerns expressed by private anglers and for-hire operators. It should involve both the MRIP communications team and the NMFS Office of Communications.

MRIP Certification Process

By developing a certification process, the MRIP made substantial progress toward implementing key relevant recommendations of the 2006 NRC report. The MRIP has invested in the development of a well-structured process for continued scientific evaluation, review, and certification of the recreational fisheries surveys. This certification process provides a framework for maintaining a national perspective for recreational fisheries data collection and for evaluating whether the regional and state efforts meet the basic MRIP requirements for stock assessment and management. Furthermore, it affords a mechanism for providing highly specialized technical and scientific support for the development, review, and certification of regional- or state-specific surveys and enhances the MRIP's ability to address regional and state needs for stock assessment and fisheries management. Although the MRIP's partners indicate they are appreciative of this increased capability and support, some are concerned about the timeliness of the review and certification process and the uncertainties associated with additional funding needs for implementation of survey improvements required for the MRIP certification.

DEGREE OF COORDINATION

The multijurisdictional nature of marine fisheries management, which in most regions of the country involves not only regional fisheries management councils but also multiple states and institutions, presents significant coordination challenges to data collection, data management, stock assessment, and ultimately fisheries management. To collect recreational fisheries data that meet required standards for assessment and management in this complex, multijurisdictional system, the MRIP surveys are conducted in cooperation with a variety of regional and state agencies as well as other institutional partners. In addition, U.S. marine recreational fisheries show wide-ranging regional differences, and in many cases differences among various fisheries within each region. These differences can be attributed to several factors, including the amount and shape of the coastline and other ocean features, species composition and diversity, and socioeconomic and demographic factors. Accommodating these regional differences requires the MRIP to adopt an implementation approach that incorporates the flexibility required to address unique regional and state needs while at the same time maintaining the standardization and national-level cohesion recommended by the NRC report.

Despite the lingering public perception of a centralized, top-down implementation approach the MRIP has been responsive to regional and state needs. Progress has been achieved in expanding and strengthening coordination and the provision of financial, logistical, and technical support to state partners, in part through regional Interstate Marine Fisheries Commissions and their associated Fisheries Information Networks and the Atlantic Coast Cooperative Statistics

Program. As a result, the program has evolved to become a compilation of regionally based data-collection programs and is better prepared to address data needs at regional and state levels.

However, challenges remain. Some state needs—for example, development of catch and effort estimates at small spatial scales for assessment or management of state-managed species, or in-season monitoring of compliance with Annual Catch Limits—have been difficult to address. This is particularly true when they require a disproportionate increase in sampling effort and become cost prohibitive, or are so specialized that they become difficult to integrate into the standard MRIP. The Pacific Coast states (Washington, Oregon, and California) are currently working with the MRIP to certify their surveys, and continued coordination, technical support, and integration of Pacific Coast state surveys into the MRIP framework are warranted. Furthermore, flat or reduced funding has made implementation of recommended survey improvements difficult. If this problem persists, advances in the states' sampling programs through the MRIP certification process potentially will be at risk.

At a regional level, increased coordination with the fishery management councils and their Scientific and Statistical Committees (SSCs) would provide increased opportunities for identifying and addressing data needs for stock assessment and management. Closer coordination with the SSCs would provide the MRIP with an additional avenue for communicating with the councils.

It is also worth noting that the timeliness of MRIP support is also dependent on capacity and funding.

COMMUNICATIONS

Overall, the MRIP has made significant advances in improving its communications and outreach strategy since the NRC's 2006 report. Perhaps its strongest advances have been with its website and its communications with some of its data-collection partners, such as the regional Interstate Marine Fishery Commissions and state fishery agencies. Its communications with some other groups, most notably anglers, but also some stock-assessment and management groups, have been less successful.

The MRIP's purpose is to estimate catch in recreational fisheries. Because fisheries management is a complex, multistage process involving many agencies and stakeholders, the MRIP should not be held responsible for explaining all facets of fisheries management.

Recommendation: NMFS should develop and lead an integrated communications strategy involving state and federal partners to explain and seek support for the management of the nation's fisheries within which the role of the MRIP is clearly defined. The MRIP Communications Plan should be an element—albeit for species in which removals are

dominated by recreational fisheries, an essential component—of such a broader, integrated overall communications plan.

The MRIP's communication efforts are guided by their Communication and Education Team and three NMFS staff (two full-time equivalents), who are tasked with the development, implementation, and coordination of the MRIP's communications strategy nationwide. The success of the MRIP depends on clear, accurate, and timely communications and on engaging all the various stakeholder groups, including anglers. Therefore, the MRIP would benefit from additional staff resources in this area.

Strategic Communications Plan

Three aspects of the 2016 MRIP Strategic Communications Plan were particularly striking to the committee. First, according to the plan, NMFS views the MRIP as a combination of state, regional, and federal efforts rather than a monolithic federal program. This is appropriate and reflects the reality that the MRIP has multiple partners who play key data-collection roles. However, this also requires a level of coordination among partners that has not been fully demonstrated.

Second, there is a lack of a needs assessment to help identify and prioritize the current communications challenges. While elements of a needs analysis are evident in other NMFS documents, an integrated, comprehensive needs analysis should be in the plan. Third, the plan lacks an implementation component, which will be essential given the challenge of reaching multiple partners and audiences. Some additional details are provided in the annual implementation plan updates on the MRIP website. However, it appears that a detailed implementation plan remains to be developed.

Audiences

This review considered three potential audiences: data-collection partners, data users, and stakeholders impacted by data, primarily anglers. The MRIP has made significant progress in expanding and strengthening the communication and coordination with regional and state data-collection partners, especially from a logistical and survey implementation point of view.

Data users include stock assessment analysts, Council SSCs and Advisory Panels, and Council and NMFS Regional Office staff who use MRIP data to implement catch limits. Assessment analysts broadly recognize the improvements in the MRIP and have found MRIP staff to be responsive to their requests for data, but would benefit from additional coordination. Engagement of the SSCs by the MRIP appears to be in the early stages and needs more emphasis. Communication to groups with responsibilities similar to those of the SSCs within the Interstate Marine Fishery Commissions and states can also be improved.

A major challenge confronting the MRIP is the use of recreational data in the management arena—specifically in implementing catch limits. Tension develops, because a survey designed for one purpose is being used for another purpose, requiring that some inferences be made. This issue was also highlighted in the 2006 report on the MRFSS. Moreover, uncertainties associated with catch estimates become critically important and may impact the timing of fishery closures. The committee also heard frustration from regional managers over the lack of timeliness of MRIP estimates for implementing catch limits—particularly when fisheries have short seasons or bursts of activity (i.e., pulses), as many recreational fisheries do. These issues can be complex and reinforce the need for an integrated communications strategy to alleviate concerns. In general, evidence presented to the committee indicated that the MRIP could be more proactive in communicating with managers and data users.

The MRIP has generally deferred communications with the anglers to the states and regions. Regional RecFIN programs and state fish and wildlife agencies conduct most of the outreach and education efforts apparently without much structured and deliberate guidance from the MRIP.

A major issue for the anglers the committee heard from was the credibility of the MRIP survey data and the data-gathering process. There are many possible reasons for their impressions, some of which can be addressed by explaining basic survey principles. Communication shortcomings have exacerbated anglers' concerns about the MRIP's value in ensuring sustainable management of recreational fish stocks. The success of the MRIP program depends on gaining the confidence of these stakeholders.

Recommendation: The MRIP should take a more active role in communicating with anglers, whether through its partners or through its own efforts. The committee recognizes that the MRIP defers to the states and regions in communications with anglers. Furthermore, the committee recognizes that an approach coordinated with the states may be most successful in building trust and aligning the understanding of these stakeholders with the reality of how the MRIP is deployed. However, the MRIP should play a leading role in providing the vision and implementation strategies that partners can follow.

For-hire captains are more likely than individual anglers to engage with the MRIP and become full partners. Currently, the MRIP has communication products aimed directly at this group, which offer direct benefits from engagement and indirect benefits from the operators' interactions with their clients.

A critical aspect of communication with all audiences, but especially anglers, is that it be a two-way dialogue. The MRIP's communication to date has focused largely on providing information. The program would benefit from greater emphasis on continually and actively collecting and incorporating feedback and input.

Strategies

Four principal strategies for communication have evolved within the MRIP. They are (1) the MRIP website, (2) the MRIP Newscast, (3) engagement of data-collection partners and data users, and (4) print and social media products.

The committee commends NMFS on the development of the MRIP website. It is well laid out, reasonably easy to navigate, and very informative. It is thorough and detailed, and NMFS appears to be developing audience-specific navigation pathways to help users find information at the appropriate level of technical detail, an improvement the committee supports. The committee applauds the transparency afforded by opportunities on the website for users to query data and view the site register. The website would be further improved if it provided an opportunity for the public to provide input.

The second mechanism for communication is the MRIP Newscast newsletter. Produced since 2008, it is a high-quality digital newsletter that provides updates and news items to recipients.

The third mechanism for communication calls for engagement of data-collection partners, data users, and others at Council and Commission meetings. Although it is appropriate to engage these audiences, they do not form a comprehensive list of audiences that should be engaged. As stated above, these effective engagements would include an opportunity to listen to stakeholder input.

The communications strategic plan proposes, as the fourth mechanism, to continue to produce both traditional and social media products that explain forthcoming changes to the MRIP, although few details were provided.

CONTINUITY

There is a need for continuity in the recreational fisheries data used for assessment, management, and allocation, because changes in time series can create challenges for management. Many important components of management are dependent on these catch and effort estimates, including stock assessment, development of harvest policies, in-season management, and catch allocation. In addition, the allocation of resources to produce catch statistics is itself dependent on the estimates of catch produced by the MRIP. The historical time series of recreational catch and effort produced with the outdated MRFSS procedures therefore requires calibration to the estimation processes used in the MRIP, so that a combined time series of total removals may be used to inform these processes.

The MRIP convened two workshops to address the calibration issues. Both workshops clearly recognized that calibration was critical in allowing stock assessments to differentiate true changes in stock status from changes in the estimation procedures producing the data used in the assessments. They also identified issues that affect the sampling error of estimates, based on changes to the survey designs over time. The workshops identified several calibration approaches, all of which invoke assumptions about effort distribution throughout a 24-hour period.

The calibrations are not straightforward because of the limited side-by-side estimation using previous and current methodology for almost all areas. The committee judges that uncertainty about process and observation error could be reduced if additional side-by-side comparisons were conducted. Continued research on calibration methodology would be useful to reducing uncertainty about stock management reference points.

Future efforts to develop calibrated time series of recreational catches will be most useful if accompanied with advice on the implications of the calibration method to stock assessment and reference points for stock management. In particular, simulation analyses of alternative methods will be helpful. As the time since a change in methodology for estimating recreational catches lengthens, the calibration method will have less influence on the understanding of current stock status. Recent data will more strongly influence stock status than will historical shifts in estimation methodology for catch. However, because the calibration methodology does influence the understanding of reference points for management, the effect of the calibration will be a persistent element of fisheries management.

Recommendation: The MRIP should continue development of a statistically sound calibration methodology as improvements to the Access Point Angler Intercept Survey and Fishing Effort Survey methodologies are incorporated. In the interim, the existing ratio-based calibration should be continued. For statistical catch-at-age (SCA)-based assessments, scientists should employ alternative catchability functions applied to the combined time series as a means to accommodate potential imprecision in the calibration of MRFSS data to MRIP data. For non-SCA frameworks, assessment scientists should exercise caution in the interpretation of trends in catch data.

1

Introduction

Over the past several decades, interest in the impact of marine recreational fishing on fish stock size and composition has increased (NRC, 1999, 2000, 2006; Lucy and Studholme, 2002; Coleman et al., 2004; Ihde et al., 2011). The recreational sector accounts for a substantial portion of the total catch in several fisheries, even exceeding the commercial catch for some species (Figure 1.1). However, several attributes of the recreational sector make it more difficult to assess and evaluate than the commercial fishing sector (NRC, 2006). This is, in large part, because there are many more recreational anglers than commercial fishermen, and the recreational sector uses a much larger number of access and landing points, on both public and private property.

THE FISHERIES MANAGEMENT CONTEXT

Because of the increasing concern about the effects of recreational fishing on fish stocks, the National Marine Fisheries Service (NMFS) of the National Oceanic and Atmospheric Administration (NOAA) has tried for more than three decades to collect and analyze data on recreational fishing. It has done this mainly through survey programs; first, through the Marine Recreational Fisheries Statistics Survey (MRFSS), and then, following a review of that program by the National Academies in 2006, through the Marine Recreational Information Program (MRIP).

Obtaining reliable data is a challenge, for the reasons mentioned above. In addition, recreational fisheries are only part of the overall fisheries management endeavor in the United States, which is a complex and multifaceted set of activities among federal, state, and joint organizations. As a result, the MRIP is not

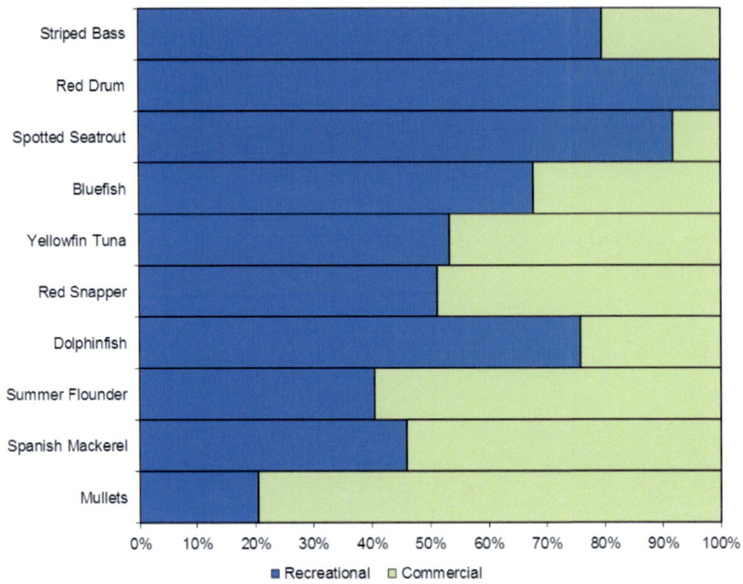

FIGURE 1.1 Comparisons of recreational harvest and commercial harvest by weight for ten popular recreational species. This figure does not include data from Alaska or Texas, which did not provide recreational weight data. SOURCE: NMFS, 2015.

implemented in a vacuum and cannot be evaluated that way. It is, and should be evaluated as, an integral part of the larger U.S. fisheries management endeavor.

To further complicate matters, the Magnuson-Stevens Fishery Conservation and Management Act (MSFCMA), the federal statute under which marine fisheries are managed, was reauthorized in 2007 with a new emphasis on avoiding overfishing and on rebuilding overfished stocks. It achieves these goals by implementing Annual Catch Limits. This changed the context of fisheries management in the United States by providing demands to limit catch, including recreational catch. As described in more detail below and in Chapter 6, this new fisheries management context changed the way that marine recreational fishery data are used. Below is a summary of the context for marine recreational fishery data (i.e., the MRIP) within the broader and more complex endeavor of fisheries management in the United States.

Federal Fisheries Management

Marine fisheries management is a complex, interdisciplinary challenge (Figure 1.2). It involves numerous stakeholders including fishers, environmental stakeholders, social and natural scientists, and managers. All agents in Figure 1.2 play essential roles by providing data, analyses, or advice and/or by implementing regulation. The figure emphasizes the involvement of recreational fisheries in the management process.

The MRIP is but one component of the fisheries management challenge depicted in Figure 1.2. The MRIP's role is to estimate recreational catch and discards of fish from the population. Discards are the fish that are released, and include those released relatively unharmed as well as those that are dead or will not survive. The total number of fish that die as a result of being caught or dis-

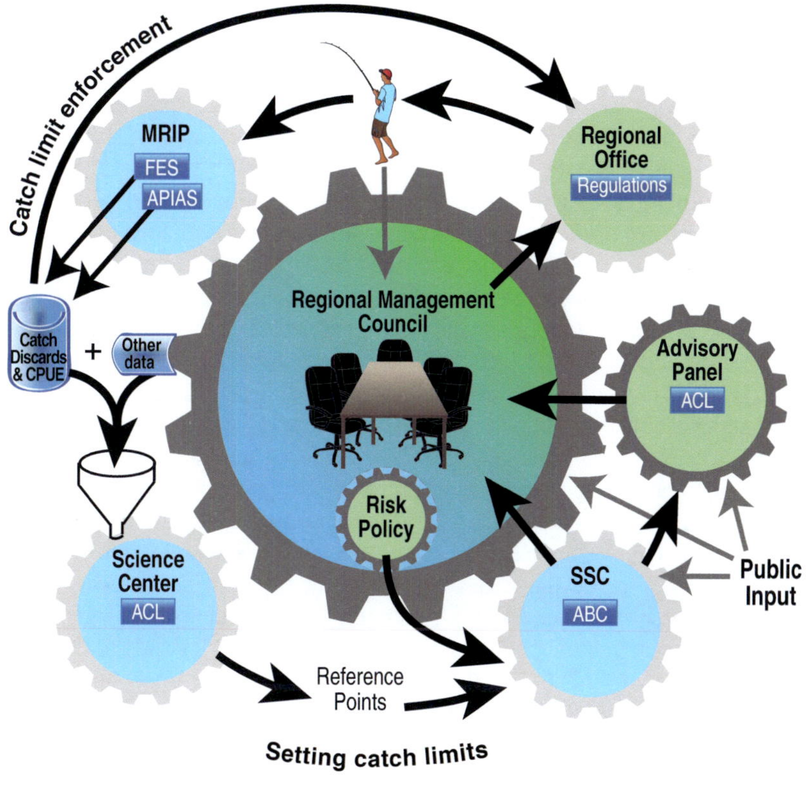

FIGURE 1.2 Schematic of the fisheries management process for recreational fisheries in federal waters. Each step of the process is represented by a separate cog in the overall system. Cogs that are primarily science based are shown as blue; those that involve societal goals are shown as green. Abbreviations used in the diagram are as follows: MRIP—Marine Recreational Fisheries Program, FES—Fishery Effort Survey, APAIS—Access Point Angler Intercept Survey, CPUE—catch per unit effort, SSC—Scientific and Statistical Committee, ABC—Acceptable Biological Catch, ACL—Annual Catch Limit. SOURCE: Committee.

carded is termed the removals. Recreational catch is estimated using statistical approaches to estimate the number of recreational angler fishing trips (effort), the average catch per trip (catch per unit effort [CPUE]), and the average number of discards per trip. The product of effort and CPUE provides an estimate of the recreational catch. The product of effort and discards per trip is weighted by an estimate of the mortality rate to estimate the total discard mortality. Additionally, CPUE in the recreational fishery is often used as an index of the abundance of the targeted species, because it is often difficult to develop reliable estimates of abundance independent of the fishery for many recreational species.

The outputs from the MRIP are used by stock assessment analysts to assess the status of the exploited fish population (Figure 1.2). A stock assessment is a mathematical representation of the population, the components of which are estimated statistically by fitting the model to observed data (Quinn and Deriso, 1999). In addition to the data from the MRIP, a stock assessment typically involves fishery-dependent data on removals (catches and discards) in commercial fisheries, data from fishery-independent surveys of abundance of the targeted species, and biological data on the targeted species. The objective of the assessment is to estimate the population abundance, fishing mortality, and stock status. The assessments are further used to determine maximum sustainable exploitation rate (when expressed as catch, this is termed the overfishing limit or OFL) and the minimum abundance that is sustainable for the species (termed the overfished limit). These estimates are reference points and are at the heart of federal fisheries management under the MSFCMA, which requires fisheries managers to avoid overfishing (i.e., not exceeding the OFL) and rebuilding stocks that are below the overfished level (MSFCMA; NMFS National Standard 1 Guidelines).[1]

The MSFCMA requires that each of the eight regional fishery management councils establish fishing policies that limit to 50 percent or lower the risk of exceeding OFL for each managed species. This is termed the council's risk policy (Figure 1.2). It is the responsibility of each council's Scientific and Statistical Committee (SSC) to use the best available science to provide a recommended Acceptable Biological Catch (ABC), which integrates the most up-to-date understanding of the status of the population of the exploited species and the Council's risk policy such that the ABC ≤ OFL to account for scientific uncertainty.[2] Each council appoints suitable qualified people, often highly trained quantitative scientists, to the SSC.

Implementation of a recommended ABC is unlikely to be perfect because of structural difficulties in regulating catch—particularly for recreational species. Accordingly, the councils are required to establish an Annual Catch Limit (ACL) such that ACL ≤ ABC ≤ OFL.[3] Councils may account for uncertainty in the implementation of their management actions by establishing an Annual Catch

[1] 16 U.S.C. §1851; 50 C.F.R. 600.310 (2009).

[2] 16 U.S.C §§1852(g)–(h).

[3] 16 U.S.C. §1852(h).

Target (ACT). For many councils, the task of establishing ACLs and ACTs is undertaken by an advisory panel composed of a diverse set of stakeholders that might include recreational fishers (Figure 1.2).

Ultimately the ACL and ACT adopted by the Council are provided to the regional NMFS office, which, acting on behalf of the Secretary of Commerce, determines the acceptability of the recommended ACL and ACT and promulgates regulations.

The MSFCMA introduced new requirements that mandate accountability measures should the ACL be exceeded.[4] For species subject to recreational fisheries, this has placed a new demand on estimates of recreational catch—to be used not only to develop OFLs, but also to ensure compliance with Council-established catch limits (Figure 1.2). The temporal and spatial demands on estimates of total annual removals for stock assessment purposes may not match the scale needed to assess when catch limits have been exceeded, requiring implementation of accountability measures.

Combined Federal and State Jurisdiction

Many species subject to recreational fishing are subject to joint federal and state jurisdictions. In such cases, a combination of appropriate agencies cooperatively manages the fisheries. However, the federal model described in Figure 1.2 is increasingly being used to manage fisheries under joint federal-state jurisdictions and even for fisheries solely under state jurisdiction. Often a single stock assessment is conducted that assumes a single, well-mixed population that is uniformly distributed throughout the region of interest, although increasingly spatially explicit models are being explored. The integrated assessment model generates a single, stock-wide ABC. These ABCs are translated through regulatory bodies into single, stock-wide ACLs and ACTs, together with regional or sector-based allocations of the ACT to each partner jurisdiction. The allocation of the ACT to regions is often based on historical patterns, and in the rapidly evolving recreational fishing sector, these allocations can be contentious (Morrison and Scott, 2014).

A separate consideration involves species managed under international governance, for example, Pacific salmon and Pacific halibut. Although the process of arriving at ACLs may differ somewhat from domestic processes, the underlying data collection and stock assessments follow science-based approaches similar to those used by U.S. agencies. However, MRIP data retain an important role in informing domestic allocations in such internationally managed stocks.

[4] 16 U.S.C. §§1853(a).

MARINE RECREATIONAL FISHERIES STATISTICS SURVEY

In 1979, NMFS established the MRFSS as a national program for obtaining standardized and comparable estimates of participation, effort, and catch within the marine recreational fisheries of the United States. The stated objective of the MRFSS was the development of a reliable national database that could be used to estimate the impact of marine recreational fishing on marine resources.[5]

The MRFSS collected data using two independent but complementary surveys, a telephone survey and an in-person intercept survey (NRC, 2006). NMFS used the telephone survey to gather information about individual anglers' fishing trips to determine the amount and types of fishing that occurred within a 2-month period, including the number, modes, access types, and dates of recreational fishing trips. The surveys inquired only about the preceding 2 months, assuming that anglers' recollections of their activities beyond 2 months were not sufficiently reliable.

The second survey used by the MRFSS was an in-person intercept survey, whereby trained field staff interviewed anglers at sites where anglers access and leave the water, such as marinas, docks, piers, or beaches (NRC, 2006). These intercept surveys were used to collect information on catch, including species, weight, length, and number of fish caught by anglers. In some cases, the onsite intercept survey was also used to collect additional biological information or samples.

Because the in-person intercept survey did not capture all anglers, and because little was known about the characteristics of the anglers sampled and those missed (to assess bias in the survey results), it was not possible to obtain a reliable estimate of total catch from the in-person intercept survey alone (Chapter 2 discusses this and other sampling issues in detail). Instead, the intercept survey was used to estimate CPUE, that is, the number of fish likely to be caught for a given unit of fishing activity. The telephone survey was required to obtain an independent estimate of angler fishing effort (E). Together, the data collected from the two surveys were used to provide estimates of total participation, effort, catch, and CPUE for six 2-month periods each year.

In addition to the intercept and telephone surveys designed and implemented by the MRFSS program, at least 13 other supplemental or component surveys were conducted by federal or state agencies to ascertain marine recreational fishery catch and effort. These additional surveys were funded at least in part through the MRFSS program and were intended to produce data that were compatible with MRFSS objectives, although the methodologies and statistical techniques often varied from the core telephone and intercept surveys conducted under the MRFSS. These additional surveys were developed as a way to better meet the data needs of a particular region or sector (NRC, 2006). Alaska has never been

[5] See http://www.st.nmfs.noaa.gov/recreational-fisheries/MRIP/program-evolution.

part of the MRFSS program, and Texas has not been a part since 1985; both conduct their own surveys.

Since the development of the MRFSS program in 1979, the context for conducting recreational fisheries surveys and the uses of survey data have changed significantly for the nation's fisheries. As exploitation levels increased, fisheries became more highly regulated, and management decisions were increasingly made at finer spatial and temporal scales (NRC, 2006; Breidt, 2013). Additionally, the mix of recreational and commercial fishing has changed over the years in many regions and for many species. By the early 2000s, some stakeholders had expressed concern that recreational data collected through the MRFSS and other recreational fishing surveys were being incorporated into management in ways that exceeded the original design and purposes. They also expressed concerns about the precision, robustness, and timeliness of the data collected through the MRFSS relative to the data needed for effective management (NRC, 2006).

2006 STUDY: REVIEW OF RECREATIONAL FISHERIES SURVEY METHODS

In 2004, NMFS requested that the National Research Council (NRC; now known as the National Academies of Sciences, Engineering, and Medicine) review data collection for marine recreational fisheries in the United States, and specifically, the MRFSS. The NRC assembled a committee of ten experts in fishery science and statistics, which released its report, *Review of Recreational Fisheries Survey Methods*, in 2006 (NRC, 2006; see Appendix B for a summary of that report). The report recommendations were categorized as sampling issues, statistical estimation issues, human dimensions, program management and support, communication and outreach, and general recommendations.

Overall, the 2006 report called for a considerable redesign of the survey program to modernize the survey methods to reduce bias, increase efficiency, and build greater trust and relationships with the recreational angling community. The report acknowledged the tremendous complexity of the challenges associated with implementing a survey program such as the MRFSS and in performing statistical analyses with the resulting data. Given these challenges, the report concluded that substantial, additional resources would be necessary to revise and improve the survey program.

THE CURRENT REVIEW

The Fishery Conservation and Management Act of 1976 has been amended and reauthorized multiple times, and is now known as the Magnuson-Stevens Fishery Conservation and Management Act. In the most recent reauthorization,[6]

[6] Magnuson-Stevens Fishery Conservation and Management Reauthorization Act of 2006, Publ. L. No. 109-479 (2007); 16 U.S.C. §§1801–1884.

Congress called for a "regionally based registry program for recreational fishermen in each of the eight fishery management regions."[7] The act further mandated that the Secretary of Commerce, "in consultation with representatives of the recreational fishing industry and experts in statistics, technology, and other appropriate fields," develop a program for making improvements in the quality and accuracy of the MRFSS.[8] The legislation particularly called for the program to implement, to the extent feasible, the recommendations of the NRC's 2006 report (see Appendixes B and C for a summary of that report and a table of its recommendations).

Since 2007, NMFS, in response to the reauthorization, has worked to improve the survey program by developing a national saltwater angler registry and transitioning from the MRFSS program to the redesigned MRIP. The redesigned MRIP includes a separate offsite Fishing Effort Survey (FES) to assess effort and an Access Point Angler Intercept Survey (APAIS) to gauge CPUE. Although the basic structure is similar to the MRFSS, major changes have been made to the methodologies and statistical analyses used for both the FES and APAIS.

The MRIP also funds and provides technical support for a variety of region-, state-, species-, and sector-specific surveys that either supplement or serve as alternatives to the APAIS and FES (Figure 1.3). NMFS has had to consider how to allow for these individual surveys, which may be better tailored for specific circumstances, while also maintaining sufficient data consistency for management.

To support this more inclusive and integrative implementation approach, the MRIP is managed via a team structure, under the guidance of an Executive Steering Committee (ESC). To ensure transparency and to achieve customer and stakeholder support, the ESC and the MRIP teams comprise members from NMFS headquarters, its regions and Fisheries Science Centers, and state agency and Interstate Marine Fishery Commissions staff. In addition, the teams are joined by participants from the regional Fishery Management Councils and key stakeholder organizations such as national recreational fishing organizations (e.g., Coastal Conservation Association). The Communications and Education Team also includes a representative from NOAA Sea Grant (Figure 1.4).

Now, a decade after the release of the 2006 report, NMFS asked the National Academies of Sciences, Engineering, and Medicine to conduct a second review to assess NMFS's progress in addressing the 2006 report recommendations. In addition, NMFS asked the Academies to consider other aspects of the survey redesign, such as the strength of the scientific process and engagement with stakeholders (see Box 1.1 for complete statement of task).

The ad hoc committee assembled to address this task was composed of nine experts in fisheries science, fisheries management, stock assessment, statistics and survey design, and social sciences. They met on four occasions, in Washington,

---

[7] 16 U.S.C. §1881(g)(1).

[8] 16 U.S.C. §1881(g)(3)(a).

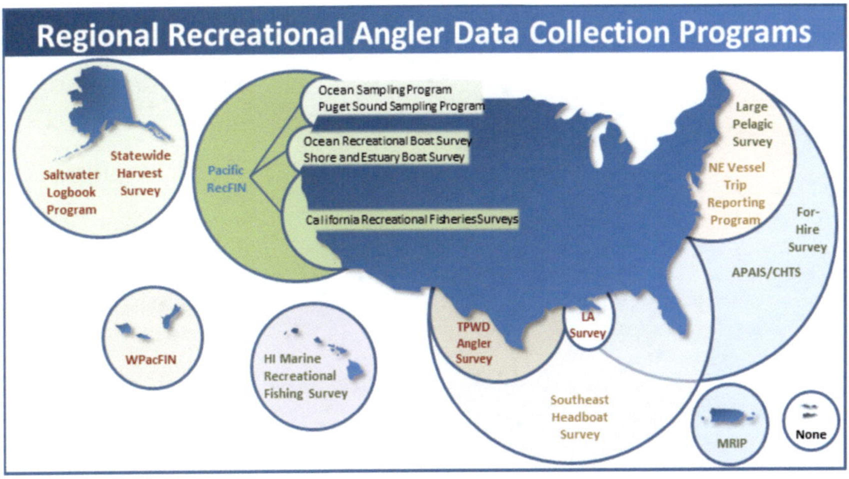

FIGURE 1.3 A visualization of where various recreational fisheries surveys are implemented within the United States. Most of the surveys are at least in part supported by the MRIP. The Texas Parks and Wildlife Department survey and both surveys conducted in Alaska, however, are not supported by MRIP funds. Represented in the individual circles (from left to right) are Alaska, Guam and Samoa, Hawaii, Puerto Rico, and the U.S. Virgin Islands. SOURCE: NMFS, 2014a.

DC (February 24-26, 2016); Charleston, South Carolina (April 25-26, 2016); New Orleans, Louisiana (May 26-28, 2016); and Irvine, California (July 11-13, 2016). At each meeting, the committee heard from representatives from federal and state government, including MRIP staff and contractors; MRIP consultants; and regional stakeholders, such as anglers, nongovernmental organizations, and representatives from fishing associations and organizations. The committee also

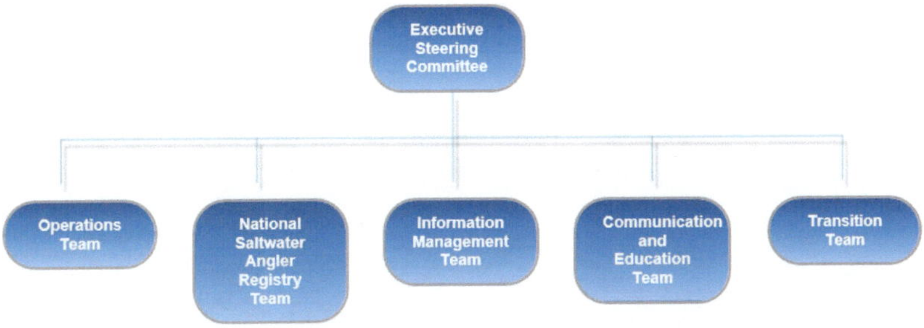

FIGURE 1.4 The team structure used to manage the MRIP. Different teams focused on various aspects of the program integrate participation of federal, regional, and state agencies and institutions. SOURCE: NOAA. See http://www.st.nmfs.noaa.gov/recreational-fisheries/MRIP/organization.

BOX 1.1
Statement of Task

An ad hoc NRC committee will assess progress in updating marine recreational fisheries data collection through the Marine Recreational Information Program (MRIP) referencing the recommendations in the 2006 NRC report *Review of Recreational Fisheries Survey Methods*. Based on this assessment, the committee will identify potential areas for improvements or changes of direction that would substantially increase data quality for fisheries management, taking into consideration potential loss of information from disruption of the time series. The committee's report will

1. Describe the approach and effectiveness of steps taken by NMFS to improve the quality and accuracy of marine recreational fisheries catch, effort, and participation statistics (in response to NRC, 2006), including, but not limited to,
   a. Establishing registries of anglers and for-hire vessels and using the registries appropriately as sample frames for recreational catch and effort surveys;
   b. Improving the effectiveness and appropriateness of sampling and estimation procedures, applicability to various kinds of management decisions, and usefulness for social and economic analyses; and
   c. Providing for ongoing technical evaluation and modification, as needed to meet emerging management needs and changes in communication technologies (e.g., smartphone apps, Internet-based social networking).
2. Assess the strength of the scientific process, including the engagement of external scientific and technical expertise, used by NMFS in developing, testing, reviewing, and certifying new sampling and estimation procedures.
3. Evaluate the communication of information on survey method development, survey method descriptions, and survey results to stakeholders and application of stakeholder input in the design and implementation of new sampling and estimation procedures. Stakeholders include at least three distinct subgroups (with some overlap among them):
   a. Data-collection partners, such as the Atlantic Coast Cooperative Statistics Program and the Fishery Information Networks;
   b. Data customers (parties that use NMFS data for stock assessments, management actions, social and economic studies); and
   c. Entities affected by the estimates (anglers and recreational fishing businesses, commercial fisheries, nonconsumptive users, etc.).
4. Determine if the degree of coordination among federal, state, and territorial survey programs is sufficient to provide a clear, national perspective on marine recreational fisheries.
5. Evaluate plans for maintaining continuity of data series to minimize disruption of management programs and stock assessments. This will include evaluation of the strategy for moving from the phone-based survey to a mail- and web-based survey as a means to estimate fishing effort.

received documents from NMFS and written input from stakeholders during the study process.

This report provides a general discussion of survey design and estimation considerations in Chapter 2. Chapters 3 and 4 provide more technical analyses of the statistical survey design and estimation procedures for the FES and APAIS, respectively. Chapter 5 discusses a framework for continued scientific evaluation, review, and certification. Chapter 6 explores the degree of coordination between the MRIP and other state and federal partners, and Chapter 7 provides an evaluation of the MRIP's communication, outreach, and education efforts. Finally, Chapter 8 reviews plans for maintaining continuity of the data series despite changing methodologies. The appendixes in this report include committee and staff biographies; the summary of the 2006 NRC report; a table of the 2006 recommendations, indicating the most relevant chapter in this report for each and the committee's ranking of NMFS responses; an excerpt from the Magnuson-Stevens Fishery Conservation and Management Reauthorization Act of 2006; copies of the survey instruments; excerpts from the 2014 Calibration Workshops; and a list of acronyms.

2

Study Design and Estimation Considerations for the MRIP

INTRODUCTION

Estimation of recreational harvest has become increasingly important with the passage of the Magnuson-Stevens Fishery Conservation and Management Reauthorization Act of 2006. Moreover, the 2006 National Research Council (NRC) report, *Review of Recreational Fisheries Survey Methods,* centered on the validity of the Marine Recreational Fisheries Statistics Survey (MRFSS) sampling for catch and effort, such as the lack of probability-based sampling in major components of the survey. In response to this and other recommendations in the 2006 NRC report, the National Marine Fisheries Service (NMFS) redesigned the survey and implemented the Marine Recreational Information Program (MRIP) to provide valid statistical estimates of recreational fisheries effort and catch. The following chapters review the Fishing Effort Survey (FES; Chapter 3) and the Access Point Angler Intercept Survey (APAIS; Chapter 4), which are two components of the MRIP, and assume technical knowledge of survey sampling concepts that may not be common knowledge. The purpose of this chapter is to provide perspective on data collection, sample design, and estimation relevant to the MRIP to help the reader who is not familiar with statistical methods for survey sampling of recreational fisheries.

CONTACT METHODS

Surveys of recreational fishing to obtain metrics of catch and effort rely on seven possible methods of contacting anglers (Pollock et al., 1994; Jones and Pollock, 2013; see Box 2.1 for discussion of the distinctions between censuses and sample surveys). Anglers can be contacted (1) onsite at public access points,

BOX 2.1
Censuses versus Sample Surveys

With surveys, the goal is to estimate finite population parameters of the population under study—most often, a population total (e.g., total catch) or a population mean. Ideally, one would conduct a census to collect information from all the units in the population. Typically, however, only a fraction of the population—a sample—is randomly selected. Reasons for conducting a sample survey rather than a census include that (1) censuses are more subject to nonsampling errors than sample surveys are, since sample surveys can often afford to allocate more resources (human and financial) to reducing nonsampling errors such as nonresponse; (2) censuses are more expensive, because they survey the entire population and data collection is an expensive step; and (3) censuses require more time to conduct, and often the data must be gathered and processed and the results disseminated within a short time frame.

In a sample survey, each population unit is assigned a strictly positive selection probability, which may vary from one unit to another. The sample is randomly selected to satisfy the selection probabilities fixed prior to sampling. The sample design consists of all the steps necessary to select a sample. The complexity of the sample design depends on the type of survey and the information available prior to sampling. Complex sampling designs usually involve stratification and clustering. Both the FES and APAIS involve fairly complex designs.

(2) by roving through the water body to seek out anglers, or (3) by aerial surveys (which capture effort only). In these situations, the field agent records the trip, its completion time, and the number of anglers and asks the anglers about the trip duration, species sought, species caught and number, and number released following a scripted questionnaire. Often, the agent can observe and measure the catch.

Alternatively, anglers can be contacted offsite (4) by telephone, (5) by mail, (6) electronically (e.g., web, email), or (7) door to door (which is rarely used). In a mail survey, the angler receives a questionnaire that asks them to report dates, trips, trip locations, and in some surveys the species and catch numbers. Questions about catch are less common, because the species and catch numbers must be remembered correctly from months past.

In measuring catch and effort, these contact methods have different strengths and weaknesses. Offsite methods obtain information that is self-reported by the angler and is not independently verified. Onsite methods, most commonly an access point survey, can verify trips and landed catches because they are observed by field agents. However, even onsite methods rely on angler self-reporting of released fish (Groves, 1989; Jones and Pollock, 2013). Released fish can be counted and verified when boats are large enough to carry an observer, such as with a headboat or charter boat. The MRIP relies on contacting anglers onsite at public-access points to obtain measures of catch per unit effort (CPUE) and offsite by mail and by telephone to obtain measures of effort. These measures are then combined to estimate total catch. This approach is used throughout the United States' Atlantic and Gulf of Mexico coasts.

Since the 2006 report was published, there have been major advances in the public's use of technology that has the potential to alter the way that surveys are done. Even though the 2006 report recommended that NMFS explore electronic reporting, the agency has only recently expanded testing of electronic reporting of for-hire logbooks, electronic capture of onsite intercepts, and web-based surveys (Kelly, 2016). For example, Liu and colleagues (2016) undertook a project to determine whether smartphone applications (apps) could be used to estimate recreational red snapper catch in Texas. Anglers reported their catches using the iSnapper app, and some of those app users were also intercepted in a probability-based onsite interview. The total catch was subsequently estimated by a modified mark-recapture method. This approach shows promise, and the committee encourages NMFS to pursue this area of research. However, self-motivated anglers who self-report via apps may not represent the target population, which presents challenges to statistical estimation—a topic discussed further in Chapter 4.

CHALLENGES WITH DATA COLLECTION

The choice of survey method is dependent on the time frame in which data are needed and the funds available to conduct the survey; both timeliness and funding issues were raised as concerns by state agencies during the committee's current review of the MRIP. Offsite methods using telephone and mail surveys are generally less expensive than onsite methods because the latter require trained personnel in the field (Groves, 1989; Jones and Pollock, 2013). Some methods, such as telephone surveys, can obtain data quickly, while mail surveys take more time. Both are complicated when the response rate is low because of the potential for nonresponse bias. Nonresponse bias occurs when respondents and nonrespondents differ with respect to the characteristics of interest (see, e.g., Lohr, 2010). For example, if people who caught fish respond while people who caught nothing do not respond because they think that their information is not needed, then the CPUE would be estimated as higher than what actually occurred. Onsite surveys can cost more per interview, but nonresponse is typically low. The use of electronic tablets for onsite surveys decreases the reporting time and, with added software, can increase data quality (Kelly, 2016).

Surveys are subject to biases beyond that of nonresponse (Groves et al., 2009; Pollock et al., 1994). Offsite surveys are subject to recall bias because of the delay between the fishing trips and receiving a questionnaire, telephone call, or electronic message. Unless anglers keep a log or diary, they may not remember trips or catch accurately. Species identification and number of fish caught can be inaccurately reported and are not verifiable. Direct biological measurements of fish necessary for estimating length and age relationships and the extraction of scales or otoliths for aging are not available. Onsite public-access surveys are subject to avidity bias (i.e., avid anglers who are better at catching fish might be overrepresented in onsite surveys) and lack coverage of anglers using private access when interviews are generally conducted at public-access sites. The lack of intercept information from most private access means that the use of CPUE requires the strong assumption that catch and effort are equal between anglers using public and private access (Ashford et al., 2010, 2011, 2013). Additionally, the error structures will differ with the type of data collection (e.g., self-administered in mail surveys versus interviewer administered in telephone surveys); this topic is discussed in Chapters 3 and 4 as it pertains to FES and APAIS, respectively. Note that there are other sources of error, such as item nonresponse in returned questionnaires, that are discussed in subsequent chapters.

SOURCES OF SURVEY ERROR

Surveys are designed to provide estimates for a possibly large number of characteristics of interest. Typically, the interest lies in estimating finite population parameters (e.g., means, percentages, ratios) of the target population, which describe some aspect of the finite population (e.g., total effort). Estimates of these population parameters are calculated from information collected on the sample, which is subject to several types of errors (Groves, 1989). The committee defines the total error of an estimate as the difference between the estimate and the true population value, the latter being unknown. The total error can be expressed as the sum of sampling and nonsampling errors (Groves et al., 2009; Biemer, 2010). Sampling errors occur because the desired information is only observed for a part (sample) of the population.

Nonsampling errors can be divided into four broad groups: (1) coverage errors, (2) nonresponse errors, (3) measurement errors, and (4) processing errors. Coverage error occurs when there is frame imperfection. This includes undercoverage (some units in the target population are not in the sampling frame) and overcoverage (some units are not in the target population but are in the sampling frame). Andrews et al. (2014) suggested that "undercoverage due to unlicensed fishing activity may be as high as 70 percent in some states for certain types of fishing activity" (see discussion of the National Saltwater Angler Registry in Chapter 3).

Nonresponse errors occur because the desired information is only observed for a part of the sample. The committee distinguishes unit nonresponse from item nonresponse. Unit nonresponse is the complete lack of information on a given sample unit. It occurs, for example, when the sampled person either is not at home when a telephone interviewer calls or refuses to participate in the survey. Item nonresponse occurs when the survey responses (items) for a sampled person (unit) are incomplete. The latter occurs, for example, because the sampled person refuses to respond to sensitive items such as fishing location or may not know the answer to some items, or because of edit failures (e.g., incorrect telephone number). Missing values may also be generated when the collected data are invalid or inconsistent.

Misresponse or measurement error occurs when the information from the respondent is inaccurate. Measurement errors can be caused by a poorly designed questionnaire or inability of the respondent to recall the requested information. Another example of measurement error is digit bias, that is, the tendency for respondents to round upward or downward (e.g., if a respondent catches 7 fish but responds with either 5 or 10), also known as rounding errors (Scholtus, 2011).

Finally, processing consists of all the handling data activities after data collection and before estimation. Processing errors occur during data coding (which is the process of assigning a numerical value to a response) and data capture.

SAMPLING FRAMES

Recreational angler surveys use sampling frames to randomly select, with known probabilities of selection, households or fishing sites and times to contact. To contact households with anglers offsite to determine effort, two approaches have been commonly used in marine fisheries (Jones, 2001). The MRFSS, the MRIP's predecessor, used a telephone survey that relied on random dialing of noncommercial telephone numbers with a coastal county prefix. In this case, the sampling frame was any household with a landline telephone number with an appropriate coastal prefix. The efficiency of random-digit-dialing telephone surveys declined over time as fewer households had landlines, more individuals switched to only having cell phones, caller ID resulted in fewer calls answered, and telephone numbers became portable. With the portability of telephone numbers, a previous coastal county resident might move inland and no longer fish. Similarly, someone from a landlocked state with that area code and prefix may move to a coastal county and become an avid angler. Using coastal county prefixes would result in both overcoverage (anglers that have moved away) and undercoverage (people that moved to the coast) of the target population. Furthermore, because surveys are subject to restrictions on dialing cell phones, the use of telephone surveys has become more problematic (AAPOR, 2016). In its 2006 report, the NRC committee recommended that alternate sampling methods be developed to address these issues of nonresponse and inefficiencies. Specifically, the report recommended that NMFS develop a national registry of all marine anglers as a sampling frame that would consist of names, telephone numbers (including cell numbers), and addresses. A sampling frame is a list from which a sample can be selected. Such a license-based sample frame would provide a targeted and efficient list for sampling.

Undercoverage for an effort survey such as the FES is managed by reweighting sampling units (e.g., angler trips) with data available from the onsite survey. The onsite survey includes all anglers, both coastal and noncoastal. The noncoastal proportion is used to expand the effort estimate of the offsite survey. This method would be unbiased if the characteristics of the people on the frame are similar to those not on the frame. Furthermore, the reliability of a survey from such a licensed-based frame declines when exemptions to license requirements are allowed, such as for retirees, military personnel, and for people under age 16. Exemptions such as these cause frame undercoverage. There is also frame overcoverage, which occurs when people on the list no longer fish. Undercoverage is likely the greater problem.

For an intercept survey, the frame consists of a list of public fishing access sites, mainly piers and boat launching sites, crossed by time of day. Returning anglers are interviewed to determine their total catch for each species. Typically, the design uses selection probabilities proportional to estimated fishing effort (e.g., site/time combinations with more effort have a higher probability of being sampled) to efficiently select the access sites for interviewing. Undercoverage occurs when access sites are excluded, and overcoverage occurs when nonmarine sites are included. When combined with an offsite survey to determine effort, undercoverage of the effort survey can be addressed using the responses to the intercept survey, by determining the percentage of anglers who were contacted but not included on the effort survey. This complemented approach is based on the assumption that anglers missed by the intercept survey (e.g., fishing on private piers) have the same average CPUE as those included on the intercept survey, which can be an incorrect assumption depending on the target species. Therefore, the assumption is made that both public- and private-access users have similar fishing patterns (Ashford et al., 2010, 2013).

INTRODUCTION TO SAMPLE ESTIMATION OF TOTAL CATCH

Although catch and effort can be expanded directly in a few situations, a widely used approach to estimate the total catch of saltwater fish by recreational anglers is to split the problem into two surveys. The importance of having these two surveys is that the FES only obtains effort in the relevant states and, because of the difficulties with self-reporting, does not obtain data on catch. Meanwhile, the APAIS obtains data on anglers with residences outside the coastal states and enables observation of catches. First, from the FES one estimates angler effort ($\hat{E}$), the total number of trips spent saltwater fishing, using a survey of all anglers within the household, asking the respondents, in a given time period, for the total number of occasions during which they have fished in saltwater either from the shore of their coastal state or from a boat that returned to the shore of their coastal state. Second, from the APAIS one estimates the catch per unit effort ($\widehat{CPUE}$), which is the number of fish caught per angler trip on each occasion, and the discard species and discard rate. If both angler effort and CPUE are well estimated for a given species, region, and period, one can calculate the total catch for that species (and region and period; see Box 2.2). Total catch (also termed total harvests in other contexts) and total discards are estimated from two surveys, one offsite and one onsite. Total discards (obtained from the onsite survey) are subsequently used to calculate dead discards (total discards times discard mortality rate). Stock assessment analysts sum the total catch and total dead discards to estimate the removals from the fish population.

Using the equations provided in Box 2.2, total removals contains the following components: (1) fish that are landed whole and can be measured and identified, (2) fish that were filleted or discarded dead, and (3) fish that were discarded alive but subsequently die from capture effects.

For stock assessment, an additional fraction of the catch released alive that subsequently dies is estimated by multiplying the fish discarded by a separately determined discard mortality rate to obtain component (3)—fish discarded alive that subsequently die from capture effects. The equations used to estimate removals, catch, effort, and CPUE by the MRIP are provided in Box 2.2.

Bias and variance of an estimator

Two important measures are usually considered when assessing the quality of an estimator: bias and variance. For simplicity, the committee assumes that estimates are only subject to sampling errors and that nonsampling errors are either negligible or adjusted for prior to estimation. By sampling errors, the committee means that each sampling event is just one possible result of many that could have occurred. In practice, only one sample is selected, but many other possible samples could have been selected from the population. Suppose that it would be possible to select all the possible samples using the same sampling design from the target population. In each sample, an estimate of the characteristic of interest (e.g., total catch) could then be computed from the observed data. The bias is then defined as the difference between the average of the estimates produced from all possible samples and the corresponding (unknown) true value for the target population. The population sampling variance is a measure of the variability of the estimates about their average that would have been observed had all possible samples been selected from the target population.

An estimator is said to be precise (or efficient) if it exhibits a small variance. Factors affecting the variance include the variance of the target population, the sampling design used to select the sample from the target population, and the sample size. For a given sample design, the variance decreases as the sample size increases. Because it refers to all the possible samples that could be selected from the population, the population variance is typically unknown but can be estimated from the selected sample. Survey statisticians use measures such as the estimated standard error (which is defined as the square root of the estimated variance) or the estimated coefficient of variation (which is defined as the ratio of the estimated standard error to the estimate). Another name for the coefficient of variation is the proportional or relative standard error.

BOX 2.2
MRIP—Estimated Removal

Population total removals

$$R = A + B_1 + (B_2 \times DMR), \quad (1)$$

where
R = removal from fish population
A = available catch (observed harvest)
B_1 = unavailable catch (unobserved harvest)
B_2 = discard
$A+B_1$ = total catch/harvest
DMR = discard mortality rate
$B_2 \times M$ = dead discard

Abbreviations

PSU = primary (first stage) sampling unit, such as site in the APAIS (see Chapter 4)
SSU = secondary (second stage) sampling unit, such as angler or boat within site in the APAIS
TSU = tertiary (third stage) sampling unit, such as angler within selected boat in the APAIS

Indexes

d = domain of interest such as state, region, fishing mode (shore, private/rental boat, for-hire), fishing area (inland, state territorial sea, federal exclusive economic zones), sampling strata
h = mutually exclusive sampling stratum by study (e.g., state by date within 2-month wave APAIS, or state by geographic distance from the shore for FES)
i = household (FES PSU)
j = site (APAIS PSU)
k = boat/angler and, if applicable, angler group within boat (see Chapter 4)

Estimated effort obtained from the FES, the Coastal Household Telephone Survey, and the like

Estimated effort = $\hat{E}$, defined as the estimated total number of single-day angler trips spent saltwater fishing

$$\hat{E} = \sum_h \sum_i w_{hi} e_{hi}, \quad (2)$$

where e_{hi} = number of angler trips per household i in state (region), stratum h (i.e., household hi)
w_{hi} = final analysis weight for household hi

Estimated catch per unit effort obtained from the APAIS and similar studies

(a) Estimated total catch per unit effort for species f for domain d =

$$\widehat{CPUE}_{df} = \frac{\sum_h \left(\sum_j \sum_k w_{hj} w_{k|hj} c_{hjk}^{(f)} \delta_{dhjk} \right)}{\sum_h \left(\sum_j \sum_k w_{hj} w_{k|hj} e_{hjk} \delta_{dhjk} \right)} \quad (3)$$

where e_{hjk} = number of angler trips per SSU hjk
$c_{hjk}^{(f)}$ = total catch ($A + B_1$) for species f per SSU hjk
δ_{hjk} = zero-one variable indicating membership in domain d
$w_{k|hj}$ = SSU-level final analysis weight conditional on selection of site hj
w_{hj} = final analysis weight for PSU hj

(b) Estimated total discard per unit effort for species f for domain d =

$$\widehat{CPUE}_{df} = \frac{\sum_h \left(\sum_j \sum_k w_{hj} w_{k|hj} d_{hjk}^{(f)} \delta_{dhjk} \right)}{\sum_h \left(\sum_j \sum_k w_{hj} w_{k|hj} e_{hjk} \delta_{dhjk} \right)} \quad (4)$$

where $d_{hjk}^{(f)}$ = discards per SSU hjk. All other symbols are defined for $\widehat{CPUE}_{df}$.

Estimated totals

Estimated ($A + B_1$), total catch for species f, is $(\hat{A} + \hat{B}_1) = \hat{E} \times \widehat{CPUE}_{df}$, where $\hat{E}$ is defined in (2) and $\widehat{CPUE}_{df}$ is defined in (3).
Estimated B_2, total discards for species f, is $\hat{B}_2 = \hat{E} \times \widehat{DPUE}_{df}$ where $\hat{E}$ is defined in (2) and $\widehat{DPUE}_{df}$ is defined in (4).
Estimated R, total removal from population, is

$$\begin{aligned} \hat{R} &= \left(\hat{A} + \hat{B}\right) + \left(\hat{B}_2 \times \widehat{DMR}\right) \\ &= \left(\hat{E} \times \widehat{CPUE}_{df}\right) + \left(\left[\hat{E} \times \widehat{DPUE}_{df}\right] \times \widehat{DMR}\right) \end{aligned}$$

where $\widehat{DMR}$ = discard mortality rate estimated via independent studies (see Chapter 4).

Assuming there is no nonsampling error, bias is generally not an issue because survey statisticians typically use unbiased (or approximately unbiased) estimators. Bias is generally caused by the presence of nonsampling errors as discussed above. Depending on the source of the bias (nonresponse error, coverage error, and measurement error), several weighting procedures are used to reduce the bias as much as possible.

Weighting Methodology

The data collected in the field are typically stored in a data file that contains rows corresponding to sample units (e.g., an angler) and columns that represent characteristics of interest (e.g., number of trips taken in the past 2 months). The file includes a column of weights that account for the sample design, coverage errors, and nonresponse, and that together constitutes a weighting system. Estimates are obtained by applying the weighting system to a characteristic of interest.

The typical weighting process consists of three major stages (see, e.g., Valliant et al., 2013). In Stage 1, each sample unit is assigned a design (or base) weight, which is defined as the inverse of its inclusion probability in the sample, a characteristic of the sampling design. Stage 2 aims to reduce the potential bias due to unit nonresponse. This bias may be large when respondents and nonrespondents differ with respect to the characteristics of interest, especially if the nonresponse rate is high. For example, the most common way to deal with unit nonresponse is to eliminate the nonrespondents from the data file and to adjust the design weights of the respondents to compensate for the elimination of the nonrespondents. To that end, the basic weights of respondents are multiplied by a nonresponse adjustment factor, which is defined as the inverse of the estimated response probability. Key to achieving an efficient bias reduction is the availability of powerful auxiliary information, which is a set of fully observed variables. Finally, in Stage 3, the weights adjusted for nonresponse are further modified so that survey weighted estimates agree with known population totals available from external sources (e.g., the census or administrative data). This process is known as calibration and can be effective at reducing the biases due to undercoverage. The resulting weights are often referred to as final weights and the corresponding weighting system as the final weighting system.

In some cases, the weighting process involves an additional stage during which the final weights undergo further modification. Most often, it consists of smoothing or trimming the weights to improve the efficiency of survey estimates. This stage is often encountered when highly variable weights are poorly related to the characteristics of interest. In such cases, the resulting estimators may exhibit a large variance (i.e., low precision). Weight trimming consists of reducing the weight values above a given threshold. These weights are set to the value of that threshold.

A point estimate for a given characteristic of interest is readily obtained by applying the final weighting system to the column corresponding to this characteristic of interest. The associated (proportional) standard error also uses the final weights but with a more complex formula than is appropriate for this report—see, for example, Wolter (2007) for additional material on variance calculations.

3

Sampling and Statistical Estimation for the Fishing Effort Survey

INTRODUCTION

Fishing effort, a key component required for the estimation of fishery removals, historically had been estimated with data collected from a random-digit-dialing (RDD) landline telephone survey within the Marine Recreational Fisheries Statistics Survey. The 2006 National Research Council (NRC) report cited a growing number of biases affecting the accuracy and precision of estimates with this study design. These included, for example, decreasing coverage of the angler population with an ever-increasing proportion of cell phone–only households, decreasing participation in telephone surveys in general, and increasing inefficiencies because of the inability to target households with one or more anglers. In response to these challenges, the National Marine Fisheries Service (NMFS) developed an innovative mail survey design through an enhanced sampling frame to improve effectiveness and appropriateness of fishing effort estimation for the Marine Recreational Information Program (MRIP).

This chapter discusses NMFS's initiatives to research and address the 2006 recommendations, along with the present committee's evaluation of those initiatives, recommendations for future pilot studies, and areas of focus to guide continuing improvements for the MRIP.

DATA COLLECTION AND SAMPLING FRAMES

The 2006 NRC report included several recommendations for improving the estimation of fishing effort, including a call for research to identify a "comprehensive, universal sampling frame with national coverage" and to address ever-decreasing response rates that limit the utility of the data. NMFS accepted this challenge and conducted a series of informative pilot studies in consultation with independent survey statisticians and survey methodologists. The committee briefly summarizes the findings presented in Andrews et al. (2014) below, beginning with relevant background on the original effort survey.

The Coastal Household Telephone Survey

The original fishing effort study, the Coastal Household Telephone Survey (CHTS), was a telephone survey conducted on a targeted random sample drawn from a list-assisted, landline RDD sampling frame. The intended population was all residents living in coastal county households identified by prespecified telephone area codes and exchanges associated with the geographic areas. The specific goal was to collect information from anglers regarding their fishing activities during the previous 2-month period (referred to as a wave). The 2006 NRC report pointed to potentially low data quality because of problems such as undercoverage bias from a growing proportion of households without a landline phone (Boyle et al., 2009; Blumberg and Luke, 2015; McCarthy, 2015), as well as already low response rates, which were projected to further decrease over time (e.g., Curtin et al., 2005; Keeter et al., 2006). Based on the experiences of several states with licenses or registries, the 2006 committee suggested that a national angler registry could provide considerable efficiencies for sampling and data collection and improved data quality over the RDD design.

National Saltwater Angler Registry

The National Oceanic and Atmospheric Administration (NOAA) established the National Saltwater Angler Registry (NSAR) on January 1, 2010 (NOAA, 2009). States with saltwater license registries were allowed to sign a memorandum of agreement (MOA) whereby their existing license frames could serve to meet the federal requirements. In accordance with the MOA, states agree to share data regarding their license holders or registrants, and in return, NMFS does not require anglers who fish in those states to register federally.[1] These states, however, had (and still have) varying exemptions from having a license. As a result, coverage of the angler population is not consistent throughout the NSAR. Table 3.1 provides a summary of coverage issues for the NSAR related to exemptions. To address the viability of using the NSAR as a sampling frame, NMFS conducted a targeted pilot study summarized below.

Several steps can be taken to address the issues of undercoverage in the sampling frame (and nonresponse to the mail survey). For example, people with license exemptions can be interviewed at access sites, and a correction factor can

[1] 500 CFR §600, Subpart P.

TABLE 3.1 Sources of Undercoverage due to the Registry Requirements of the NSAR by State[a]

State	Basic License Requirement	Exception for Fishing on Licensed For-Hire Vessel?	Exception for Anglers 16 and Under?	Other Major Exceptions?
Alabama	Salt Water License	Y	Y	N. License-exempt residents over age 64 and anglers fishing on state-licensed piers must obtain a state saltwater angling registration.
Alaska	Fishing License	N	Y	N. Residents over age 60 are exempt, but must obtain a Senior Alaska Resident card
California	Fishing License	N	Y	Fishing on a public fishing pier.
Connecticut	Salt Water License	Y	Y	N
Delaware	Fishing License + FIN # Registration. Anglers declare intent to fish in salt water on FIN registration.	Y	Y	N (Persons exempted from license requirement must still have FIN #)
Florida	Salt Water License	Y	Y	Residents over age 65. Anglers fishing from state-licensed piers.
Georgia	Fishing License + Saltwater Information Program (SIP) registration	Y	Y	N
Hawaii	None	-	-	-
Louisiana	Salt Water License	N	Y	Persons who turned 60 yrs of age before 6/1/00 are exempt. Persons who turned 60 after 6/1/00 must have a senior license.

continued

TABLE 3.1 Coontinued

State	Basic License Requirement	Exception for Fishing on Licensed For-Hire Vessel?	Exception for Anglers 16 and Under?	Other Major Exceptions?
Maine	Saltwater fishing registration OR Freshwater License + state they fished in saltwater	Y	Y	N
Maryland	Bay and Coastal Fishing License OR Bay Boat License OR Bay and Coastal Registration if exempt (e.g. unlicensed angler on licensed boat)	Y	Y	Fishing from commercial pier. Piers provide list of users to registry. Persons otherwise exempted from license must obtain registration.
Massachusetts	Salt Water Permit	Y	Y	N
Mississippi	Salt Water License	N	Y	N
National Saltwater Angler Registry[b]	Registration	Y	Y	Any person currently licensed by, or a resident exempted from the state's license requirements by, an Exempted State; persons angling for non-anadromous species in state waters.
New Hampshire	Salt Water License	Y	Y	N
New Jersey	Salt Water Registration	Y	Y	N
New York	Salt Water Registration	Y	Y	N
North Carolina	Coastal Recreational Fishing License	Y	Y	Grandfathered lifetime license holders as of 1/1/2006; anglers fishing on licensed piers
Oregon	Fishing License	N	Y (under 12)	N

Pennsylvania	Fishing License + Pennsylvania Saltwater Angler Registration (lower Delaware River only)	N	Y	N
Rhode Island	Salt Water License	Y	Y	N
South Carolina	Salt Water License; Senior Fishing License	Y	Y	Anglers fishing on licensed piers.
Texas	Fishing License + a Saltwater Stamp	N	Y (under 17)	Persons born before 1/1/1930 are exempt.
Virginia	Tidal Waters Fishing License OR Boat License OR Fishing Information Program (FIP) Registration if exempt (e.g., fishing on licensed boat)	Y	Y	N. Anglers exempt from licensing must get FIP registration.
Washington	Fishing License	N	Y (under 15)	N

[a] https://www.countmyfish.noaa.gov/register.

[b] Per www.countmyfish.noaa.gov/register: "Starting January 1, 2011, if you have a saltwater recreational fishing license or registration from any state or U.S. territory EXCEPT Hawaii, Puerto Rico, or the U.S. Virgin Islands, you are AUTOMATICALLY registered and do not need to take further action." (NMFS Communication to 2016 committee, 22 June 2016.)

SOURCE: Modified from NMFS.

be applied to the Fishing Effort Survey (FES). Also, survey personnel can try to contact people who do not return a mail survey by email or telephone.

Pilot Study—Angler License Directory Telephone Survey

NMFS conducted pilot studies in several states with established angler registries, including the Angler License Directory Survey (ALDS). The data-collection methodology for the ALDS was similar to the CHTS as both were telephone surveys, but its sampling methodology was different. CHTS samples were drawn from randomly generated telephone numbers in designated coastal area codes and exchanges but no household information (e.g., name, angler license status) was available. ALDS samples, by contrast, were randomly selected from the licensure database with associated contact information that was sometimes incomplete or out of date. Thus, the pilot study afforded a direct comparison of the two sampling frames—landline RDD for the CHTS and the angler registry for the ALDS. Although the ALDS response rates were only "marginally higher" than those for the CHTS, the new sampling frame resulted in significant data-collection efficiencies through an increased number of interviews from the target population of saltwater anglers (Andrews et al., 2014). This research, however, also suggested that sizable coverage issues existed with the registries related to errors in contact information (e.g., old/incorrect telephone numbers), state-specific exemptions, and anglers that should have a license but did not, ranging as high as 70 percent in some states.

Pilot Study—Dual-Frame Telephone Survey

In addition to the angler registry, the 2006 report also suggested a general dual-frame approach to increase coverage of the target population for estimation of fishing effort (Lohr, 2007; Brick et al., 2011) and to increase data-collection efficiencies with already identified anglers, regardless of the chosen data-collection mode (e.g., telephone, mail). Building on the results of the ALDS pilot, NMFS examined the combination of the angler registry with a landline RDD survey. Acknowledging that anglers could be listed in either or both sampling frames, NMFS selected independent samples from each frame (CHTS RDD and ALDS registry) and then weighted the results from each frame to produce a series of unified estimates (e.g., Lohr and Rao, 2000). Response rates from the pilot survey were low and similar in magnitude to those for the CHTS; undercoverage concerns noted for the ALDS design remained. In addition, there was insufficient information to determine whether a sampled household/angler was listed in both frames to construct efficient weighting adjustments to lower nonresponse bias and coverage bias associated with those not present in either frame. Consequently, NMFS abandoned this alternative design.

NMFS evaluated neither a landline/cellular dual-frame RDD design for the estimation of fishing effort, nor a telephone survey using cell phone numbers alone. Instead, it assessed prior research on telephone surveys for characteristics relevant to the needs of the MRIP. For example, since May 2004, residents may "port" their landline (or wired) numbers to a cellular (wireless) carrier and device (FCC, 2016). In addition, cell phone numbers can travel, meaning that the number assigned upon activation need not change when a person moves to another area in the United States. Thus, ported-landline and cell phone numbers introduce inefficiencies in data collection because they are not necessarily linked to the geographic areas targeted by the MRIP (e.g., Keeter et al., 2015).

Fishing Effort Survey

The ALDS research uncovered sampling and data-collection efficiencies in using the NSAR as a sampling frame (as suggested in the 2006 report), but NMFS noted that the remaining undercoverage could limit the quality of the fishing effort estimates. Additionally, the general, ongoing decline of response rates to telephone surveys was a growing challenge. For example, Brick et al. (2011) discovered that the coverage of the CHTS was only about 50 percent in the aggregate of Florida, Massachusetts, New York, and North Carolina, and that the aggregate response rate was around 10 percent, while a test mail survey resulted in a response rate of greater than 30 percent. Consequently, NMFS evaluated the feasibility of a mail survey.

Address-based sampling (ABS) frames have been available to the public since the early 1990s (Iannacchione, 2011). These frames are developed from commercially available versions of the U.S. Postal Service's Computerized Delivery Sequence (CDS) file, the route taken by postal carriers to deliver mail. The CDS, like the NSAR, alone is not a complete list and is therefore subject to undercoverage. The CDS may be supplemented with information to produce a more complete sampling frame. Supplemental files include, for example, the No-Stat file, a file containing more than seven million primarily rural mailing addresses not listed on the CDS (Shook-Sa et al., 2013), and ancillary data from public and private sources related to population demographics and other characteristics (AAPOR, 2016). With augmentation of the No-Stat file alone, ABS frames provide near-complete coverage of the U.S. household population (Iannacchione, 2011).

NMFS then tested a new list that incorporated the coverage benefits of ABS and a state-specific licensure database (NSAR) for the new FES. All ABS addresses in relevant East and Gulf Coast states were retained, excluding grouped quarters without individual unit addresses (e.g., correctional and nursing facilities; Reist, 2012) and known businesses. Additional records found on the NSAR that did not match information on the ABS address list were also retained, including those with addresses outside the coastal state. Addresses on the new FES sampling frame were then stratified (grouped) within state to allow for differential sampling by (1) coastal counties (specified distance from the coast) versus noncoastal counties and (2) NSAR exact match (address and/or telephone number, if available) versus no match (Andrews et al., 2014). The sampling literature refers to this design as a single-stage stratified design (e.g., Valliant et al., 2013).

The FES pilot test using the new address frame resulted in impressive improvements over its predecessor survey, the CHTS (Andrews et al., 2014). The augmented-ABS frame enabled a direct link to coastal households through geolocation information; this provided efficient sampling and data-collection methods to target angler households. In addition, this new approach provided a new level of stratification for sampling associated with licensure status (Yes versus No/Unknown). Then, samples in the matched strata were drawn at a higher rate to gain efficiency under the assumption that this stratum has higher rates of saltwater anglers. Finally, a subsample of nonrespondents was contacted to assess nonresponse bias; data collected did not show any detectable levels of nonresponse bias, suggesting high-quality data. Many studies include nonresponse follow-up components to their study designs to measure and adjust for nonresponse bias following guidance provided in *Standards and Guidelines for Statistical Surveys* from the U.S. Office of Management and Budget (OMB, 2006).

FES documentation to date is not clear on the level of augmentation of the sampling frame beyond the NSAR such as the No-Stats file. This suggests an area of future research toward ensuring maximal coverage of the coastal-state household population especially for those with private boat docks. Additional augmentation of the FES frame could afford further targeted sampling and associated data-collection efficiencies if information, for example, from market research vendors proves fruitful.

The 2006 report recommended that a dual-frame survey (i.e., using more than one sampling frame to draw a probability sample) "should be used wherever possible to reduce sample bias" associated with undercoverage noted for the single sampling frame design (see Chapter 2 discussion). Although not a true dual-frame design as suggested by the 2006 report, NMFS correctly argues that the advantage of its approach is that it avoids biases in the dual-frame estimator resulting from identification of households listed on multiple frames (Andrews et al., 2014). These records can be identified either prior to sample selection through frame matching or post data collection through respondent-provided information such as whether they have a saltwater fishing license. Here, the frame matching errors create modest efficiency loss but do not create bias since the weights remain the inverse of the probability of selection and all households are covered by the frame.

Results from the FES pilot study were striking. The new study design produced a 1.6-fold increase in the likelihood of surveying a household with at least one angler over the other pilot designs evaluated. There was also a threefold increase in the response rate, along with a 4.1-fold increase in "the mail survey estimate of total fishing effort" relative to the CHTS (Andrews et al., 2014). The sizable increase in estimated effort from this pilot does not necessarily suggest higher-quality data (e.g., lower nonresponse bias), but it could indicate a true change over time; without a "gold standard" (from another survey or source) on which to compare, the reason for the change is only speculative.

NMFS officially launched the FES in January 2015 in tandem with the CHTS for states in the Atlantic and Gulf Coast regions (NMFS, 2016a). Administering the two surveys simultaneously is part of the 3-year plan to transition from the CHTS to FES while gathering needed information to recalibrate historical CHTS estimates to adjust the data series. The FES achieved an overall response rate of 35.1 percent (ranging from 32.3 percent to 44.7 percent), almost 28 percentage points higher than the CHTS (7.3 percent, ranging from 4.6 percent to 11.2 percent). As with the pilot study, fishing effort estimated from the FES was 4.7 times larger than the value tabulated from the CHTS responses.

Other Data-Collection Research

The 2006 NRC report also mentions the need to investigate other modes such as electronic data collection. Although much has been accomplished, to our knowledge NMFS has yet to investigate the role of electronic data collection (e.g., a web-based survey) either alone or in combination with an initial mode of data collection. This investigation, however, should proceed with caution. For example, providing participating households with access to a web instrument (in lieu of completing the mail questionnaire) may provide cost savings and reduce the time needed to key and process the data. However, the digital divide (Horrigan, 2015) may result in coverage bias by excluding lower-income households, which would suggest that this option is not viable for full implementation. Data collection via a smartphone application (app) or text messaging may supplement the web option in a mixed-mode survey as long as the questionnaire remains relatively short (Link et al., 2014); coverage bias is less likely a concern here because an estimated 92 percent of adults in the United States have a cell phone (Anderson, 2015).

However, mixed-mode surveys, that is, those with multiple ways for a respondent to provide information (e.g., mail a hardcopy questionnaire or use the web), present several advantages and disadvantages (Dillman and Messer, 2010). Advantages include reduction of errors associated with non-negligible nonresponse and possibly reduction of survey costs. Disadvantages include mode effects (differential patterns of reported information associated with the data-collection methods), lower data quality (Sakshaug et al., 2010), and possibly lower participation rates (Medway and Fulton, 2012).

Fishing Effort Estimates from the For-Hire Surveys

The CHTS includes only households with a landline area code and exchange linked to coastal counties of the United States. The FES sampling frame is much larger, consisting of all addresses in the East and Gulf Coast states. Unlike the pilot study, the FES sampling frame excludes anglers identified from licensure databases who live outside these coastal states. Consequently, undercoverage in the FES frame will remain if this list does not adequately cover noncoastal state anglers. Because the MRIP's scope covers recreational angler fishing effort regardless of where the person lives, both the CHTS and FES may include some level of undercoverage in the fishing effort estimates if the adjustment for non-coastal anglers estimated from the Access Point Angler Intercept Survey (APAIS) is somehow insufficient (Fisheries Statistics Division, 2016). The committee welcomes ongoing analyses of FES coverage, both before and after the APAIS adjustment is applied, along with the direct evaluations of the APAIS noncoastal adjustments.

The For-Hire Survey (FHS) was designed to collect information on "fishing effort and catch by marine recreational anglers fishing on professionally licensed for-hire vessels (including charter, guide, and large party boats)" simultaneously (Sauls et al., 2008). The FHS was initially "developed to resolve undercoverage of charter and headboat angler effort" inherent in the CHTS for the Atlantic and Gulf Coasts (NMFS, 2014b). The committee presumes that the FHS may also provide an undercoverage adjustment for the FES to either confirm or supplement the APAIS adjustment. NMFS states in the MRIP Data Users Handbook that most anglers who take these types of boat trips do not live in coastal states (NMFS, 2014a).

Unlike the CHTS and FES, the FHS includes samples of for-hire vessels selected from a "comprehensive directory of for-hire boats" stratified by vessel type, state, and week within the data-collection wave. To date, the committee is unaware of studies to assess and address the coverage properties of the FHS sampling frame and agrees with a consultants' report that stresses the need for a comprehensive list (Chromy et al., 2009). The handbook notes that an adjustment factor from the APAIS is applied to the FHS effort estimates to account for angler trips on for-hire vessels not on the sampling frame. Details of the undercoverage adjustment and other survey weight components are found in Sauls et al. (2008). Evaluative studies along with documentation on sampling, frame coverage, and other measurement issues for the FHS would benefit the MRIP and provide needed information to the public.

A vessel representative is contacted by telephone to relay details of the fishing trips that occurred during the prior week, including the number of customers who fished for a particular period. The committee noted that the survey does not ask respondents to identify the number of anglers living outside the coastal areas, or whether they have their own fishing license (and hence are captured on the NSAR). Thus, the committee cannot confirm or refine the statement that most anglers on Atlantic/Gulf Coast for-hire vessels are from noncoastal areas. Gathering such information may be feasible from a cost and burden perspective if NMFS collects electronic logbook information from the vessel captains as recommended by Chromy et al. (2009). Additional information on the FHS is discussed in Chapter 4.

SURVEY MATERIALS

The 2006 report did not provide key recommendations for the study questionnaire used to estimate effort. However, noting that interviewer-assisted questionnaires are different from self-administered ones (Groves et al., 2009; Dillman et al., 2014), both in form and in content, NMFS set out to develop and test a new questionnaire (also called the survey instrument) for the FES.

NMFS used cognitive testing (Groves et al., 2009) to evaluate changes in the short CHTS instrument to improve, for example, the angler's ability to report on saltwater fishing sites to the exclusion of freshwater sites. NMFS also focused on telescoping errors where respondents inadvertently include or exclude fishing trips from the designated 2-month reporting period (Gaskell et al., 2000). Pilot studies conducted by NMFS suggest that these challenges have been reduced, although they recognize that some "residual reporting errors" may still exist.

As with the CHTS questionnaire, the FES questionnaire is relatively short, covering both sides of one page. The FES questionnaire contains 10 questions on weather information, whether anyone in the household has been fresh- or saltwater fishing in the past 12 months, the type of telephone service, household tenure (e.g., rent, own), length of stay, and household size. Nonfishing questions are included in the FES questionnaire based on research that shows such items increase participation from non-angler households (NMFS, 2014a). Six questions are asked of, at most, the five oldest members of the household: demographics, whether they saltwater fished from shore/boat, and the number of days fished by location in the designated 2-month period and within the past 12 months. Because of the undercoverage of private-access anglers in the intercept survey, an additional question to determine whether respondents used public or private access would provide valuable information to the MRIP.

Materials included with the mail questionnaire are a cover letter providing details about informed consent to participate in FES, along with frequently asked questions, a prepaid return envelope, and a small cash incentive. Noting challenges with respondents distinguishing between fresh- from saltwater locations, NMFS could evaluate the utility of including a state map with identified saltwater access points. This may also improve angler recall.

Another item of note is the 2-month recall period common to both the CHTS and FES. Limited documentation is available on the historical decision to set 2 months as the recall period (Groves et al., 2009; see also Chapter 1) for CHTS other than methodological studies conducted in the 1970s that suggested a recall period longer than 2 months would result in unreliable estimates (NMFS, 2014a). Presentations from NMFS to the committee in open meetings suggest that reducing the size of the data-collection wave and the recall period could increase the FES sample size and consequently the cost of the study. NMFS will soon comment on results from a recent pilot experiment to compare results from a 1-month recall period against the standard 2-month period. Any design changes to address non-negligible recall bias should be made in light of sample size (cost) implications, along with effects on when estimates are made available to the public.

Noting problems associated with recall bias, research has been conducted using prospective data-collection techniques. For example, the Migratory Bird Hunter Survey requests sample members to maintain a prospective diary to record hunting trips during the season (USFWS, 2016). However, some research suggests that prospective diaries could increase participant burden and lower response rates, and should be evaluated through a pilot study (Fricker and Tourangeau, 2010).

Prospective electronic data submission by a household respondent, perhaps through smartphone or tablet apps, may ease these concerns and could be a focus for future research. The ability to "capture data in the moment" may reduce recall bias, an issue raised about the CHTS and FES, provided that the participation burden does not affect participation rates (Link et al., 2014). Therefore, NMFS is encouraged to consider a prospective design with electronic data collection as a future pilot study. As discussed in Chapter 8, NMFS should consider implications on the data series when evaluating the pros and cons and introducing enhancements to the FES.

SAMPLE DESIGN

The CHTS sample design is described as a stratified simple random sample of RDD landline telephone numbers associated with targeted coastal areas and subareas. The study telephone numbers are randomly selected from banks of 100 numbers with at least one working residential landline phone (1+ banks), excluding those designated as a business (Link et al., 2008). Biases associated with undercoverage (excluding, for example, cell phone–only households) and cognitive burden in recalling fishing trips in the past 2 months during a brief telephone interview are a few challenges noted in the 2006 report and by NMFS (Andrews et al., 2014).

The FES sample design, by contrast, is a stratified simple random sample selected bimonthly from an ABS frame of addresses in Atlantic and Gulf Coast states. Mutually exclusive strata (groups of addresses) are defined by the interaction of county proximity to the coast and NSAR match (yes/no) within each state under the FES purview, all important characteristics to the estimation of the annual fishing effort. NMFS uses differential sampling rates to target strata with a higher likelihood of interviewing anglers without sacrificing coverage

(Andrews et al., 2014). Just prior to data collection, NMFS augments the sample with state-specific license registry data linked to the NSAR to ensure current contact information such as telephone number (Fisheries Statistics Division, 2016). Documentation on the augmentation could benefit from additional clarity regarding the point at which the frame is updated with critical information such as the NSAR match (yes/no; see also Chapter 7 recommendation regarding enhanced documentation).

The FES is designed to produce cross-sectional (i.e., yearly) fishing effort estimates by state. As noted in documentation provided to this committee by NMFS, the state-level annual estimates are expected to be precise, assuming a coefficient of variation (= 100 × standard error of the fishing effort estimate divided by the estimate) no greater than 20 percent and historical response rates (Fisheries Statistics Division, 2016). An optimal allocation methodology determines the distribution of cases across strata within each state. Requiring the FES to produce precise estimates for in-season estimation is not feasible given time and funding constraints. Doing so would require specialized surveys for this purpose—consider, for example, the red snapper survey field tests being conducted in Alabama, Florida, Mississippi, and Texas in collaboration with NMFS (Sharpe, 2016)—and/or specialized statistical methodology.

The 2006 report recommended the evaluation of panel designs for estimation of effort, noting both pros and cons of this alternative design. With survey panel designs, all or a portion of the sample is interviewed across multiple data-collection periods (e.g., Lavrakas, 2008). The 2006 committee focused specifically on the benefits of a rotating panel design, whereby change between 2 years of the study can be estimated along with cross-sectional changes as currently implemented. Also mentioned in the 2006 report were the potential benefits of a rotating panel design (multiple panels of sample members are brought in and removed from the study at a designated frequency) for maintaining or even increasing response rates (e.g., Lavrakas, 2008). To date, panel pilot studies have been conducted in Texas to evaluate the utility of the iSnapper app for prospective collection of catch data on red snapper (Stunz et al., 2014; NOAA, 2016a) and in North Carolina and Florida to assess the feasibility of collecting catch and effort simultaneously (NOAA, 2016a).

NMFS is cognizant of bias associated with nonresponse and has included design components to mitigate this challenge. All FES sampled households that do not respond within a specified time period receive a reminder postcard. The third and final mailing to remaining nonrespondents includes a nonresponse conversion letter, a second questionnaire, and a postage-paid return envelope, delivered together via first-class mail. In addition, the pilot studies that served in the development of the current FES design included an evaluation of nonresponse bias (Groves and Couper, 1998). NMFS conducted a small nonresponse follow-up study on a random subsample of nonrespondents (see, for example, Valliant et al., 2013 for a discussion of nonresponse follow-up studies). This subsample was contacted again using priority mail and an additional cash incentive. Comparison of the "fishing prevalence" estimates did not uncover substantial differences in the initial and follow-up respondents. However, the documentation does not discuss changes in fishing effort. NMFS should evaluate the utility of including a nonresponse follow-up as part of the standard FES design as an ongoing evaluation of nonresponse bias.

WEIGHTING

Historically, fishing effort was estimated with the CHTS as the weighted number of saltwater fishing trips made by coastal-area residents inflated to include an estimate for noncoastal area residents tabulated from the APAIS (Chapter 4). With the FES, nonresident anglers—those on the NSAR and without a corresponding address on the ABS frame—contribute data for the nonresident estimates, while those on the ABS frame provide the core effort estimate. These surveys are referred to as the FES Nonresident Angler Survey (NAS) and the FES Resident Angler Survey (RAS) in some documentation (NMFS, 2013).

Regardless of the survey, the base weights (inverse selection probabilities) are adjusted for nonresponse to mitigate biases potentially present in the respondent data if those who decline to participate have differing levels of effort (e.g., Lohr, 2010; Valliant et al., 2013). NMFS uses a nonresponse weighting class adjustment with classes formed with information available for all sampled households, namely, the interaction of state, coastal/noncoastal area, and additionally with the FES, match/nonmatch with the NSAR, and presence of a telephone number linked to the sampled ABS address. Survey weights are generated independently for RAS and NAS. With this methodology, respondents and nonrespondents ideally respond similarly for key study questions within groups formed by the weighting classes (Valliant et al., 2013; Haziza and Lesage, 2016). This is a strong assumption made for all surveys using this approach.

As noted previously, the FES ABS frame appears to contain only information appended from the NSAR and no other source. NMFS may find that a nonresponse model enhanced with NSAR information could prove of benefit for the matched sample if item nonresponse and data quality were sufficient to warrant investigation. Additional model covariates may be obtained through supplemental information provided on the ABS frame (e.g., indicator for a seasonal home) or market research vendors (AAPOR, 2016).

In the final step, NMFS calibrates the nonresponse-adjusted weights for the study respondents to the estimated number of households by substate sampling strata from the American Community Survey (ACS; Fisheries Statistics Division, 2016). Not only does this procedure align the estimated number of households with the ACS, but also weight calibration has been shown to lower both sampling and nonsampling errors if relevant variables are available for respondents and from the population (Kott and Chang, 2010; Kott, 2016). Data obtained through the FES questionnaire, such as household tenure, may prove advantageous to enhance the calibration model. In addition, as noted in the 2006 report, a rotating-panel survey could afford detailed variables for nonresponse adjustment for panel members who participated in the first year of the study but not the second.

In summary, FES weighting methodology includes three key components: inverse probability of selection, an adjustment for nonresponse, and poststratification. Documentation to date does not suggest any treatment for mail packets returned as undeliverable or weight adjustments for ineligibility (e.g., vacant households). NMFS could consider a separate unknown-eligibility adjustment especially if the proportion of the sample with no contact is large. Otherwise, the fishing effort estimates could be overinflated because the weighting methodology must assume the same rates of recreational fishing for unoccupied households as calculated from responding households. We assume that the population control totals do not include unoccupied households and therefore address this issue. However, enhanced documentation on the weighting methodology would benefit NMFS now and in the future, as well as provide additional information for the public at large.

Additionally, if NMFS further expands the bimonthly design to include a nonresponse follow-up, then further research is needed to evaluate the weighting methodology in light of a two-phase design, where phase 1 is the current FES design and phase 2 is the nonresponse follow-up. For example, correlates of nonresponse may differ by phase, suggesting a different nonresponse adjustment for the follow-up study. Enhancing the complexity of the design and/or the weighting methodology must be carefully evaluated to determine relative gains in efficiency and data quality without delaying release of the estimates or affecting continuity of the data series (see Chapter 8).

NMFS also could use the FES to estimate the number of households with at least one angler in U.S. coastal states. If these FES estimates do not align with the population, then estimated effort could be severely biased low or high. Consider this generic example: Unbeknownst to the research team, the FES sampling frame had 25 percent undercoverage of the angler population, a conservative estimate given the 70 percent result cited in Andrews et al. (2014). A higher proportion of sample addresses was drawn from the NSAR-matched cases in keeping with the current design. In keeping with leverage-saliency theory—where people who are interested in the survey topic are more likely to participate in the survey (Groves et al., 2000)—the response rate from angler households was 2.5 times higher than from non-angler households. For convenience and simplicity, we ignore the effect of measurement error in the data provided by the participating households. Instead, we focus on the final weight calibration step noted above. If the base weights are adjusted to the frame totals, then the estimated number of angler trips (effort) could be underestimated because of the undercoverage bias. Conversely, if the base weights are adjusted to the ACS totals, then the estimated number of angler trips could be overestimated because of nonresponse bias.

As demonstrated through this simple example, comparison of the estimated number of households from the FES with other sources is very important. However, to our knowledge, such information is nonexistent. Consequently, NMFS should consider collaborating with other federal agencies to include items on their surveys to estimate the number of recreational saltwater anglers or households with at least one angler, such as the American Time Use Survey. These external estimates could be used to verify the estimates or as covariates in an FES weight calibration model to reduce nonsampling biases (Dever and Valliant, 2010).

DATA QUALITY AND MISSING DATA

Many components define data quality, including coverage, nonresponse (both item and unit), questionnaire content, data entry, and sampling error. Biemer and Lyberg (2003) provide a framework for assessing quality through the lens of total survey error (see also Chapter 2 discussion). NMFS has made great strides in redesigning the effort survey to lower bias and improve data quality. However, the assessment of data quality must be ongoing, such as by including a nonresponse follow-up as a standard component of the FES design.

Other potential issues include respondent compensation, respondent perceptions, and the validity of retrospective data. The wave follow-up methods in use (e.g., reminder postcard) appear adequate and fit within the framework of standard mail-out survey methodologies. Based on the findings of an MRIP-sponsored pilot study, NMFS determined the optimal compensation to surveyed households to be $2.00 (Andrews et al., 2014). The findings of the study appear reasonable, and the choice of $2.00 reflects a careful consideration of the trade-offs between nonresponse reduction and survey cost. Another potential problem is respondent perceptions. Respondents' perceptions of government, the value of the MRIP survey, and the effectiveness of management efforts at various levels may vary by state. Variations in response rates across states, considered when determining the final sample size for each state, should be monitored and assessed on a regular basis. For states with particularly low response rates, efforts should be made to research underlying reasons (e.g., insufficient incentive), perhaps through a nonresponse follow-up, and to develop appropriate strategies to mitigate the problem.

Although discussed earlier, concerns about the validity of retrospective data certainly require further scrutiny. At least one pilot project is reviewing the measurement error and validity of the 2-month reference for estimating effort in the mail-out surveys. Potential measurement error is certainly one problem. However, the problem is somewhat more complicated by the fact that one person from each FES household is likely reporting on the fishing efforts of the other members of the household, similar to the CHTS design. There is the potential for measurement error when the respondent reports his or her effort, as well as when they report other household members' effort, for the same 2-month time period. Thus, measurement error in this case can take various forms.

One area of data quality not covered in the current FES documentation is item nonresponse, that is, missing responses to some questions from an otherwise completed questionnaire (Haziza, 2009). Item nonresponse can be addressed with imputation, where the missing value is replaced with a valid response using a defined model. Conversely, missing values can be excluded from the household-level estimates; in the case of fishing effort, this assumes no effort (fishing trips) for one or more household members. NMFS is encouraged to report on the level of item nonresponse and to identify procedures to address the incomplete information, because it has direct implications on estimation. Methods could include weighted hot-deck imputation (Cox, 1980) with predefined classes for quick implementation, or more advanced techniques for questions with high item nonresponse or increased likelihood for rounding bias (e.g., Huttenlocher et al., 1990).

VARIANCE ESTIMATION

A standard error for estimated effort is calculated through Taylor expansion procedures per information provided to the committee by NMFS (Wolter, 2007; Fisheries Statistics Division, 2016). The Taylor expansion approach does not account for nonresponse unlike other methods such as replicate variance estimation (e.g., Valliant, 2004). If the sampling fraction (i.e., proportion of households selected for the study out of those on the FES sampling frame) is small, the so-called reverse approach of Shao and Steel (1999) is also another option (see also Haziza, 2009; Kim and Rao, 2009). Software is available to analyze both sets of weights. However, the generation of replicate weights requires both additional time and additional research to determine how many replicate weights to generate (e.g., Wolter, 2007; Valliant et al., 2008).

CONCLUSIONS AND RECOMMENDATIONS

Conclusion: The methodologies associated with the current Fishing Effort Survey, including the address-based sampling mail survey design, are major improvements over the original Coastal Household Telephone Survey that employed random-digit dialing to contact anglers. This is a reflection of an immense amount of effort on the parts of the NMFS staff, contractors, and consultants.

Conclusion: The 2-month recall period for the Coastal Household Telephone Survey (CHTS) was set to be consistent with the seasonal time periods captured by the onsite intercept surveys, such as the Access Point Angler Intercept Survey. This same recall period was chosen for the Fishing Effort Survey to match the CHTS. Several factors, however, are related to the quality of angler recollections, including total number of fishing trips and the frequency of trips around the beginning/end of the data-collection wave.

Recommendation: NMFS should continue to evaluate the cognitive properties of a 2-month recall period to confirm or update the research on this topic conducted in the 1970s.

Recommendation: NMFS should consider evaluating a prospective data-collection methodology, such as asking people in advance to document fishing trips planned over the next 2 months, to reduce concerns about angler recall.

Conclusion: Survey material initially sent to the sampled household includes a small cash incentive in appreciation of the adult respondent's time to complete the questionnaire. Incentives have been shown to be effective in reducing nonresponse. Nonresponse, however, will be an ongoing challenge for all surveys, which can lower quality and precision.

Recommendation: NMFS should consider conducting targeted annual nonresponse studies as a standard component of the MRIP. The purpose of these studies would be to continually monitor correlates of nonresponse and nonresponse bias to control its damaging effects on data quality.

Conclusion: Maintaining comparability across the years is important for evaluating trends in fishing effort. Changes in fishing effort can result from actual change over time. They can also result from measurement errors such as nonresponse bias, from procedural changes such as new survey questions, or from ineffective adjustments to the survey weights. Without data on respondents who are repeatedly surveyed over the time period of interest, it can be difficult to determine the extent to which a change is real or resulting from these other sources.

Recommendation: As recommended in the 2006 report, NMFS is encouraged to continue research on survey panels, where a portion of the sampled households is retained for one or more interviews, for the Fishing Effort Survey alone or for an effort-catch combined study. The purpose of the survey panel would be to assess trends and any anomalies in those trends, to assess any improvements in data-collection efficiency through increased participation, and possibly to lower measurement error associated with, for example, trip recall with a more engaged sample of anglers.

Recommendation: NMFS should evaluate the benefits of collaborating with another federal survey (e.g., the American Time Use Survey) to include items related to fishing effort. These external estimates could provide corroboration of the fishing effort estimates, as well as useful variables for an enhanced Fishing Effort Survey weight calibration model to address sampling and nonsampling biases.

Conclusion: Collecting data for fishing effort estimates through electronic modes (e.g., web questionnaire, smartphone app) may reduce study costs associated with keying and processing the questionnaires. In addition, these vehicles may be a viable option to increase release of fishing effort estimates with data that are evaluated in real time.

Recommendation: As recommended in the 2006 report, electronic data collection should be further evaluated as an option for the Fishing Effort Survey, including smartphone apps, electronic diaries for prospective data collection, and a web option for all or just panel members.

Conclusion: Weight adjustments have proven effective in lowering biases in survey estimates such as those associated with nonresponse and frame coverage errors. The effectiveness is only as great as the association of the adjustment covariates with nonresponse and important measures of the survey. The Fishing Effort Survey weighting methodology borrows on the strength of the new sampling design to include, for example, an indicator for at least one licensed angler in the household. Consequently, the use of additional variables that are associated with fishing effort and/or survey participation might prove beneficial for the weight adjustment models.

Recommendation: Current or augmented variables on the address-based sampling frame should be evaluated to improve the efficiency of the Fishing Effort Survey weighting methodology.

Conclusion: Variance estimation is a critical component to any survey. Methods that do not account for all components of the sampling design and weight adjustments will typically underestimate the sampling variance. This is especially important for surveys without a high level of response such as the Fishing Effort Survey (~40 percent).

Recommendation: Other variance estimation methods should be evaluated for fishing effort estimates to account for weight adjustments, especially those associated with nonresponse. These include replication methods and the so-called reverse approach.

4

Sampling and Statistical Estimation for the Angler Intercept Survey

INTRODUCTION

One important component of recreational fishing surveys is the intercept survey. As noted by the National Research Council's (NRC's) 2006 study, the intercept survey for the Marine Recreational Fisheries Statistics Survey (MRFSS) used a stratified multistage sampling design, but the (point and variance) estimation procedures did not account for the complex design features, potentially leading to biased estimates. Also, the previous MRFSS design did not adequately cover night fishing. Finally, that design used the concept of "alternate site," which did not support the calculation of well-defined inclusion probabilities. To address the 2006 report recommendations, the intercept survey underwent a complete redesign in terms of both sampling and estimation procedures. The current methods used in the Access Point Angler Intercept Survey (APAIS) for the Marine Recreational Information Program (MRIP) are a vast improvement over the previous sampling and estimation procedures and reflect state-of-the-art methods in survey sampling.

This chapter discusses the initiatives implemented to address the recommendations, the current committee's evaluations of those initiatives, and recommendations for future studies and improvements.

DATA COLLECTION AND SAMPLING DESIGN

The target population for the intercept survey consists of the marine recreational angling fishing trips that are taken during a given 2-month data-collection period, or wave. For these purposes, a "trip" is generally considered to be each time an angler engages in fishing and then subsequently leaves a particular site.

Between 5 million and 20 million fishing trips are estimated to occur during a given wave. The newer APAIS is conducted for 2-month waves in 16 states bordering the Atlantic Coast and Gulf of Mexico (excluding Texas and Louisiana), as well as in Puerto Rico (Breidt and Chromy, 2016). Depending on the state, the number of waves ranges from three to six.[1] The objective of the intercept survey is to estimate the catch per unit effort (CPUE), by species, catch category (harvested or released dead or alive), and fishing mode (charter boat, party boat, private or rental boat, shore fishing) of anglers participating in marine recreational fishing in the study states.

The two main data-collection tasks of the APAIS are counts of completed angler fishing trips and angler-intercept interviews. The angler interviews are obtained by intercepting marine recreational anglers at shore (SH), private/rental boat (PR), and charter boat (CH) access points. Sampling in the party (or head) boat (HB) mode includes riding on the boats during fishing days (no overnight fishing trips are sampled). The interviewers collect demographic data for the anglers and ask anglers about their fishing day, including any fish released or already filleted. The interviewer also examines the catch for species identification and enumeration and may weigh and measure the catch.

The current APAIS sample design is a multistage stratified design. The population is first stratified based on site group (beach-bank, artificial structures, charter boat, private/rental boat, and the special offshore group), state, wave, region, month, type of day (weekday and weekend), and 6-hour blocks within a 24-hour day (2AM-8AM, 8AM-2PM, 2PM-8PM, 8PM-2AM, 11AM-5PM). The 11AM to 5PM interval, which corresponds with peak fishing activity, was added in 2014, due to lower activity in the early morning and late afternoon/evening resulting in a small number of completed interviews (Breidt and Chromy, 2016). The inclusion probabilities are adjusted to account for the overlap with the time intervals 8AM to 2PM and 2PM to 8PM to avoid double counting. This interval corresponds to peak fishing activity. Before 2014, the APAIS used the fishing mode as a stratification variable. In 2014, the APAIS transitioned from fishing mode–stratified sampling assignments to mixed boat–mode and shore-mode sampling, and in 2016 it transitioned to fully mixed-mode sampling to ensure adequate sampling in all modes of eligible fishing anglers (Breidt and Chromy, 2016). The current APAIS design uses the site group as a stratification variable to ensure sufficient sample size for all modes of eligible fishing anglers. This decision was made to improve productivity in terms of number of completed interviews.

The first-stage sampling frame is from a spatiotemporal list of site-days (defined as a combination of a fishing site or cluster of sites and a day), which is constructed from the public-access fishing site register (SR). Field observations are entered into a web application upon return from the field. If a site closes per-

[1] See http://www.st.nmfs.noaa.gov/recreational-fisheries/Surveys/coverage.

manently (e.g., out of business, destroyed by a storm), it is retired but remains on the SR. A primary sampling unit (PSU) is a site-day within a given 6-hour time slice stratum (Breidt and Chromy, 2016). The SR is a database of all known public-access sites from Maine to Mississippi and Puerto Rico with fishing activity. Each site on the SR is assigned an identification number, which remains unchanged over time. The SR is updated regularly by field observation. The site status is coded as retired, making the site ineligible for sampling.

Each PSU consists of either a single site or two sites. Each site is assigned a fishing pressure that corresponds to a prediction of the mean number of angler fishing trips that an assigned interviewer would encounter based on the site's most common form of fishing (e.g., shore fishing, charter boat; see Table 4.1). The assigned pressure for a given site is time-interval dependent. That is, a given site may be labeled as a high-pressure site for one interval block (e.g., 11AM-5PM) and labeled as low pressure for another interval block (e.g., 8PM-2AM). Only sites with a pressure of "3" or less can be clustered with one additional site as long as the driving time between them is less than 60 minutes and they are located in the same county. These are referred to as "two-site assignments." Undercoverage is an important issue in the current APAIS, because the first-stage sampling frame contains almost exclusively public-access sites. Thus, private sites cannot be selected in the sample as they have zero probability of inclusion. If the proportion of private sites is large and the behavior of private sites differs from that of public sites in terms of catch, estimators of total catch (see Chapter 2) may suffer from large biases (e.g., Särndal et al., 1992; Särndal and Lundström, 2005).

TABLE 4.1 Pressure Category and Corresponding Size Measure

Pressure Category	Expected Range of Number of Angler Trips	Size Measure Assigned to Pressure Category
0	1-4	0.5
1	5-8	2.5
2	9-12	9
3	13-19	13
4	20-29	20
5	30-49	30
6	50-79	50
7	80+	80
8	Unable to determine	0
9	Mode not present at site or inactive sites	0

SOURCE: Breidt et al., 2012.

At the first stage, a sample of site-days is selected from the spatiotemporal list according to a probability proportional-to-size (PPS) sampling design. That is, the inclusion probability of a given site is proportional to its fishing pressure. Table 4.1 shows the pressure and associated size measure used in the PPS procedure. The latter is expected to lead to efficient estimates if the PSU inclusion probabilities are (approximately) proportional to the PSU catch. However, PPS procedures tend to be vulnerable to the presence of influential units when this "proportional relationship" is not satisfied. In other words, influential units may be the high-pressure sites with low catch or low-pressure sites with high catch. These units tend to make the classical estimators unstable in the sense that they have a large variance. One way to check whether or not the PPS procedure is appropriate is to plot the pressure measure against the total estimated catch for each site-hour combination and to check if the relationship is linear and goes through the origin. If this plot is not approximately linear through zero, then using the pressure measure as an indication of fishing activity may result in inefficient estimates.

To meet operational constraints (e.g., interviewers are not available on the selected dates), a rejective sampling procedure (Fuller, 2009) was developed, whereby a very large number of samples is first selected through PPS sampling, and only those samples satisfying the operational constraints are retained. Then, a sample is selected through simple random sampling of the samples that satisfy the operational constraints. The inclusion probability of a PSU at the first stage is then approximated through Monte Carlo methods (Breidt and Chromy, 2016); that is, the inclusion probability of a PSU was obtained as the proportions of samples that contained that PSU among the samples satisfying the operational constraints. For two-site assignments, the inclusion probabilities are adjusted to account for the overlap so that double counting does not occur. Without additional information on the sample allocation used at the first stage, the committee was not able to assess its effect on the efficiency of the estimates.

Depending on the type of fishing (shore or boat), there are one or two additional stages of sampling. Sampling of shore fishing is based on a two-stage sampling design, where the secondary sampling unit (SSU) is an angler trip within each PSU. Sampling of boat fishing is based on a three-stage sampling design, where the second stage consists of selecting boat trips within a selected site-day and the third stage consists of selecting angler groups within each boat trip selected at the second stage. The angler groups are the tertiary sampling units (TSUs). When possible, field staff try to achieve a census within a PSU. That is, on a given site-day, all the anglers present on the site are interviewed. However, a census is generally not possible because of refusals, language barriers, and missed eligible participants (see below). A census is never possible for a two-site cluster because the interviewers, intercepted SSUs, and intercepted TSUs are treated as if they were selected by simple random sampling without replacement at the second and third stages, even if this is not the case in practice. Therefore, the validity of the estimates in terms of bias depends on how well simple random sampling without replacement serves as an approximation of the actual but unknown design.

An assignment consists of a time interval, a cluster of fishing sites with activity in at least one mode, the order in which those sites are to be visited (in the case of a two-site cluster), and the date and time when the cluster will be visited. Assignments are sampled, and field staff are assigned to a date/time assignment.

The APAIS instrument is relatively short (NOAA, 2016b). In the previous MRFSS design, the field sampling procedures provided survey staff with considerable flexibility. In addition, interviewers could obtain interviews from alternate fishing modes and/or sites to increase productivity and minimize the survey costs (see Breidt et al., 2012). As a result, the inclusion probabilities were difficult to compute, making design-based type estimation a difficult task. As recommended in the 2006 report, the current design does not allow samplers to decide where and when to conduct an interview. Instead, each assignment corresponds to a fixed time interval. For two-site assignments, the sampler is told the order of sites—making it relatively straightforward to compute the inclusion probabilities that will be used in the estimation procedures. Following the recommendation of the 2006 report, the alternate sites have been eliminated. Finally, unlike in the previous MRFSS design, there is no upper limit on the number of interviews. Samplers attempt to obtain the largest possible number of completed interviews for a given assignment. Over the years, the number of completed interviews has varied from approximately 6,800 to 25,800 for a given wave in the states where the APAIS is conducted. Sampling in the HB mode includes riding on the boats during fishing days. The interviewers collect demographic data for the anglers and ask anglers about their fishing day, including any fish released or already filleted. The interviewer also examines the catch for species identification and enumeration and may weigh and measure the catch. For at-sea sampling on headboats, the interviewer remains on the boat throughout the trip, collecting data on the catch as long as fishing continues.

The APAIS uses face-to-face interviews, which enables the interviewer to clarify unclear questions and to gain the respondent's confidence. There will be variations in interviewer effects due to interviewers' experience, training, and skills. These factors may affect the nonresponse and measurement errors.

The responsibility for training interviewers is shared by the National Marine Fisheries Service (NMFS) and its data-collection partners: Atlantic Coastal Cooperative Statistics Program, Gulf States Marine Fisheries Commission, and the Atlantic and Gulf state agencies. NMFS must approve all training programs. However, it seems that NMFS has limited control on the actual implementation of interviewer training and testing as it is currently conducted. Details about the training program can be found in the APAIS Statement of Work (2016) and Chapter 5 of this report.

In addition to variation in training, various other interviewer effects and interviewer-related variance affect data quality. Face-to-face interviews, in contrast to mail surveys, can introduce additional sources of error and bias through interviewer and interviewer-respondent interactions. Some interviewers are more skilled than others, for example, which can impact inter-interviewer variance—a potential source of nonsampling error. Inter-interviewer differences can be related to similarities and differences (e.g., ethnicity) between interviewers and respondents. Thus, certain characteristics of the interviewer may affect the willingness of the angler to be interviewed or respond to certain questions—although likely not to a significant extent with the APAIS given the nonsensitive nature of its questions.

The interviewer effect will vary based on interviewer experience, training, and adherence to protocols. Experience will affect interviewer confidence, proper survey pace, and methods for gaining access to respondents. Also, experience can inform interviewers' abilities to build rapport with respondents and respond to questions about the MRIP in a positive and informative manner, which is particularly important given the confusion and distrust that exists among some constituents. Furthermore, some interviewers are more skilled than others at asking questions, which influences interviewer error and variance. Interviewers may vary in their abilities to probe deeper during open-ended questioning. For example, question 12 of the North Carolina APAIS asks anglers whether most of their fishing effort on the current trip took place mostly in the Atlantic Ocean or "other." Interviewers are expected to code "other" responses according to Department of Marine Fisheries waterbody codes, which requires both knowledge on the part of respondents and familiarity on the part of interviewers to elicit valid and reliable responses.

Positive interactions between interviewers and respondents are critical not only for collecting high-quality data, but also for promoting the program and ensuring angler participation. Anglers talk to other anglers and will share their experiences with the program via their informal social networks and social media. The goodwill arising from a positive experience will yield future benefits. The attitudes of interviewers will impact interview success in terms of unit nonresponse, item nonresponse, measurement error, and ultimately inter-interviewer variance.

Of course, interviewer behavior and how it is perceived and interpreted is important. For example, interviewers may be perceived as idle—and wasting taxpayer money—if they have extra time between interviews or are assigned to a site during bad weather when no or few anglers are present. Interviewers could mitigate possible misperceptions by engaging in alternative activities during idle periods, such as cleaning the area or conducting qualitative interviews with available anglers. In addition, the interviewer's appearance contributes to perceptions. Dress, the outward display of government symbols (e.g., government symbols on shirt), and the display of official government identification will all influence the interview outcome. Professionalism is important, but its degree and nature should be tailored to the state and regional setting in which the interviews take place. Although not a true interviewer effect, weather (e.g., temperature) can affect a respondent's willingness to be interviewed; some studies have found that high temperatures are associated with higher nonresponse rates and higher measurement error (Cohen and Krueger, 2016).

MISSING DATA

APAIS estimates are subject to missing data. For example, in the Atlantic and the Gulf of Mexico, approximately 20 percent of data are missed because not all eligible anglers complete the interviews (Breidt and Chromy, 2016). Four sources that lead to missing data are (1) refusals, which occur when an angler or party refuses to participate in the survey; (2) mid-interview refusals, which occur when the angler or party does not answer some key questions; (3) language barriers that occur when the interviewer and the angler speak different languages; and (4) missed eligibles, which occur when interviewers are busy with other anglers or parties (NMFS Statement of Work, Access Point Angler Intercept Survey). NMFS believes that the most common language barrier situation arises when a Spanish-speaking angler is approached by an English-only speaking interviewer. If an agent is too busy to interview all anglers, the selection of anglers can introduce bias if anglers are selected by party type (e.g., large groups or non-English speaking anglers are avoided). Bias can be avoided if there is no selection according to party type and if all anglers that were not interviewed are counted. A different situation exists when anglers refuse an interview, because this can be indicative of other issues (e.g., illegal catches). The committee is not aware of the percentage of missing data attributable to each of these four reasons.

Interviewers should attempt to collect some paradata, which are variables about the data-collection process (e.g., Kreuter, 2013). Paradata may include variables such as the number of anglers in the party, their gender and approximate age, and the fishing mode. These variables may be incorporated into the estimation procedures, which may help reduce the potential bias due to missing data.

WEIGHTING AND ESTIMATION

Prior to 2006, the estimation procedures did not account for the features of the complex survey design. As noted in the 2006 report, the validity (in terms of bias) of the estimates relied heavily on some (implicit) model assumptions. With the new sampling design, the inclusion probabilities and the sampling weights are well defined, making the use of design-based type procedures possible. The weighting process is therefore completely new since the 2006 report. The base weights at the first stage are defined as the inverse of the cluster inclusion probabilities. For a cluster consisting of a single site, the base weight of the site is equal to the base weight of the cluster. For a cluster consisting of two sites, a weighting methodology was developed accounting for the duration of the visit of each site. When a census is possible the weight of an angler group is equal to the weight of the cluster. When a second stage and a third stage of sampling are involved, the weights are equal to those that would have been assigned had the SSU and the TSU been selected according to simple random sampling without replacement. Thus, the validity of estimators in terms of bias depends on whether or not this assumption holds. Based on the weighting system described above, an estimate of *CPUE* (defined as the estimated number of fish caught or discarded per angler on a single saltwater fishing trip) is readily obtained, where the *CPUE* is defined as the ratio of the estimated *Catch* to the estimated *Effort,* and *Effort* is the total number of single-day angler trips spent saltwater fishing (see Chapter 2). The CPUE, produced by the APAIS, is computed at the state/fishing mode/wave/fishing area level.

An estimate of the variance of the point estimates is obtained through Taylor expansion procedures by assuming that the clusters at the first stage are selected with replacement. In practice, the clusters are selected without replacement within stratum. As a result, it is anticipated that the variance estimators will exhibit an upward bias (slightly larger than necessary), although the affect is likely to be small if the first-stage sampling fraction is small (i.e., the percentage of PSUs selected from the total available for the study is small [Särndal et al., 1992; Wolter, 2007]). In the APAIS, the first-stage sampling fraction is small, because the total number of PSUs in the population is very large in comparison to the number of selected PSUs. Therefore, it is expected that the variance estimator will perform well in terms of bias.

DISCARD ESTIMATION

NMFS estimates that a substantial quantity of recreational catch is discarded rather than retained and landed. Discards include fish released relatively unharmed and those that are dead or will not survive. In 2014, approximately 60 percent of the national recreational catch was discarded before landing due to regulation or angler choice (NMFS, 2015). Of that total, approximately 63 percent occurred in the Southeast region of the United States (Table 4.2).

TABLE 4.2 Estimated Total Recreational Catch and Percentage Released at Sea for the Entire United States and the Southeast Region, 2014

AREA	Total Catch (number of fish in thousands)	Harvested	Released	Percent Released
Nationally	392,285	155,248	237,037	60%
Southeast Region (SER)	248,797	96,866	151,931	61%
SER Relative Contribution	63%	62%	64%	

SOURCE: MRIP presentation to Fishery Management Council Coordination Committee, 2015.

TABLE 4.3 Estimated Recreational Harvest and Discards of Red Snapper in the Gulf of Mexico Region, 2010-2015

Estimate Status	Year	Region	Total Harvest (A + B1)	Released Alive (B2)
FINAL	2010	SOUTH ATLANTIC	62	102,867
FINAL	2010	GULF OF MEXICO	333,689	1,435,847
FINAL	2011	SOUTH ATLANTIC	1,049	56,455
FINAL	2011	GULF OF MEXICO	520,269	1,521,243
FINAL	2012	SOUTH ATLANTIC	7,148	106,454
FINAL	2012	GULF OF MEXICO	590,804	1,425,044
FINAL	2012	CARIBBEAN	3,842	0
FINAL	2013	SOUTH ATLANTIC	18,393	83,507
FINAL	2013	GULF OF MEXICO	1,241,780	2,824,058
FINAL	2014	SOUTH ATLANTIC	88,817	285,306
FINAL	2014	GULF OF MEXICO	391,079	1,786,360
FINAL	2014	CARIBBEAN	39,914	31
FINAL	2015	SOUTH ATLANTIC	1,111	508,196
FINAL	2015	GULF OF MEXICO	584,008	1,542,998
FINAL	2015	CARIBBEAN	34,685	125

SOURCE: MRIP data query, July 9, 2016.

An understanding of the magnitude of the discard issue for an individual species basis can be gained using data from red snapper recreational fisheries in the Gulf of Mexico region (Table 4.3). For this fishery, the percentage of the total catch discarded at sea is estimated to range from 0 percent to 99 percent, but there is little verification of the estimated quantities of discarded fish. This leads to a somewhat illusory precision in these estimated quantities. Estimated discard mortality associated with these fisheries varies by region, depth, and release method. For recreational red snapper fisheries, estimated discard mortality rates range from 10 percent to 22 percent, meaning that mortality from discarding at sea rivals that from removals landed by recreational anglers.

From a programmatic perspective, the primary elements needed for assessment and management are the number or biomass of fish that are caught and landed and the number or biomass of those that are caught and released but which subsequently die due to capture effects. For fisheries management purposes, the mortality of these discarded fish is either assumed or estimated via a discard mortality rate (DMR) and subsequently converted to biomass using an estimated weight for the discarded fish (usually an average weight). The cumulative sum of these elements is the total mortality required for assessment purposes; in MRIP terms, A + B1 + DMR(B2).

The MRIP estimates the number of fish released in the recreational fisheries (B2) in several stages, depending on region. In almost all regions, the primary method of estimating the number of discarded fish is through the APAIS, and the basic data are either self-reported by anglers or reported in mandatory logbooks in the Northeast Region for-hire program and in the Southeast Region charter boat logbook program. Attempts to validate angler reports of the quantities and size composition of discarded fish have been limited; thus, the estimates of discarded numbers and sizes are subject to considerable uncertainty (Benaka et al., 2014). There is also considerable uncertainty about the estimated mortality of these discarded fish. Although not uniquely responsible for determining discard mortality, the MRIP is responsible for determining discard rates. A major review of current knowledge, ongoing research, and data gaps conducted by NMFS in 2013 (Benaka et al., 2014) catalogued the mortality estimates currently used in fisheries management by species and regions. For recreational fisheries, the mortality estimates range from 0 percent to 100 percent, depending on species, region, and research basis for the mortality of released fish. These estimates can be strong and influential assumptions for both stock assessment and fisheries management. The problem of unknown or highly uncertain estimates of both the quantity of discarded fish and the discard mortality rate for many species is common throughout the United States. The committee notes that research into the correct DMRs for use in fisheries assessment and management is not the responsibility of the MRIP, but the committee also notes that close coordination between the MRIP and the agencies that are directly responsible for estimating DMRs for recreational fishery releases would be valuable to the stock assessment process.

The issue of total mortality estimation, including discard mortality, was also noted in NRC (2006). Although significant improvements in the APAIS have been made in the MRIP, uncertainty in the estimation of mortality and biomass of fish discarded in recreational fisheries remains a hurdle for management of many species. Because the quantities of discarded fish that die can be of similar magnitude to those that are landed for many species, the implications of this uncertainty for both the determination and management of Annual Catch Limits are profound. Uncertainty and the possible downward bias in the estimates of total mortality for species that is associated with the current framework will result in underestimates of the underlying productivity of stocks and misspecification of reference points for fisheries management. Management of Annual Catch Limits may also be compromised by inaccurate accounting of total removals.

Estimation of total mortality can be improved through

- Validation of angler-/logbook-reported discard quantities
- Comprehensive estimation of the size composition of discarded fish
- Additional research on and application of DMRs for released fish
- Additional research on statistical properties of discard mortality estimates

The committee understands that the MRIP is not solely responsible for all of these efforts. The MRIP should be concerned with the validation of discarded numbers and their size composition because they directly affect the ultimate estimates of total removals. The calculation of DMRs is a broader responsibility of fisheries management. Recognizing the importance of the discard mortality issue, the NMFS undertook a national process to develop an Action Plan for fish release mortality science in 2015 (Benaka et al., 2016). The broad goals of this action plan were to

1. Support the use of planning tools such as the SMART (Benaka et al., 2016) tool to help managers, scientists, and other stakeholders determine which fish species, complexes, and/or fisheries would benefit most from improved mortality rate estimates. The SMART tool assesses and scores the impact of discarding species based on several criteria: restricted or rare status, vulnerability to exploitation, economic impact, political sensitivity and stakeholder engagement, and discard ratio.
2. Facilitate the development of improved mortality rate estimates.
3. Support effective and efficient research that leads to reduced release mortality for high-priority species, complexes, and/or fisheries.
4. Ensure that improved mortality rate estimates are incorporated effectively into existing management processes.

These objectives pertain in part to the MRIP programs of discard estimation, technology evaluation, research funding, and communication/outreach with the angler community. Therefore, there is considerable opportunity within the MRIP for improvements in discard mortality estimation and its use in stock assessment.

The mortality resulting from the discard of recreationally caught fish is the ultimate metric of importance to stock assessment and management. The production of more accurate estimates depends on not only a more comprehensive understanding of discard mortality from future research initiatives, some within NMFS and some within academic institutions, but also involves MRIP activities and mandates. We consider the primary MRIP responsibility to be the production of reliable estimates of the number and size composition of discarded fish. Estimation of the mortality associated with these discards will require coordinated research on DMRs with other components of NMFS and partner agencies. However, verification of self-reported discards is an important role for the MRIP, and additional efforts using electronic and human-based observations are required. For example, electronic monitoring could provide a cost-effective and less intrusive option for verification than human observers, and directly addresses the responsibility of the MRIP to estimate quantities of discards rather than quantities and DMR.

COMBINING THE FISHING EFFORT SURVEY (FES) AND THE APAIS

Estimates of *Effort* in the FES and *Total Catch* in the APAIS are obtained by combining both surveys. On the one hand, the FES estimates of *Effort* are based solely on in-state residents and do not account for out-of-state residents. On the other hand, the APAIS collects information on angler trips for both in-state and out-of-state residents. As a result, it is possible to determine an estimate of the proportion of out-of-state residents from the APAIS. This estimated proportion is then used to correct the estimated *Effort* produced by the FES.

An estimate of *Total Catch* is obtained as

$$\textit{Total Catch} = \textit{Effort} \times \textit{CPUE},$$

where *Effort* is the total number of single-day recreational angler trips spent saltwater fishing and is imported from FES, and *CPUE* is the estimated number of fish caught or discarded per angler on a single saltwater fishing trip. As mentioned above, the *CPUE*, produced by the APAIS, is computed at the state/fishing mode/wave/fishing area level. Therefore, the estimator of *Total Catch* can be viewed as a stratified ratio estimator, with estimation of effort from the FES multiplied by CPUE (number of fish caught from APAIS divided by the APAIS estimation of effort). The estimated *Effort* from the FES is not expected to suffer from significant undercoverage because of the new sampling frame and adjustments for out-of-state angler trips; therefore, it is hoped that this estimator will help to ameliorate the problem caused by the absence of private sites on the first-stage sampling frame for the APAIS (e.g., Särndal and Lundström, 2005). The underlying assumption is that private sites and public-access sites share the same behavior in terms of CPUE. Although unlikely to be observed in practice, it is not currently clear whether this assumption is a critical source of bias. This assumption needs to be studied. The variance of *Total Catch* is estimated through Taylor expansion procedures taking into account the fact that the *Effort* produced by the FES is an estimate (Goodman, 1960). NMFS has developed a computer program that allows data users to obtain domain estimates for any domain of interest. However, if the domain is too small (i.e., with a very sample size), then the point estimates may not be reliable. In this case, small area estimation techniques may prove useful for finer domains. The committee raised some concern that some data users may be using this program for fine-scale domains (which are those exhibiting a small sample size), resulting in unreliable estimates.

FOR-HIRE SURVEYS

Prior to release of the 2006 NRC report, NMFS recognized difficulties in obtaining catch and effort data from the for-hire fishery. The for-hire fishery consists of several components: guide boats, small charter boats often called six-packs, and larger charter and headboats. How these components are effectively sampled depends on whether they are moored at specific ports and on their hours of egress and ingress. Smaller vessels, such as guide boats, can use public launches or docks and may not operate on rigid schedules. Their catches are intended to be captured in the APAIS. Licensed vessels are sampled in the For-Hire Survey (FHS). Unlicensed vessels are intended to be identified during the onsite surveys. Larger vessels that are located at specific sites and operate on defined schedules (e.g., leave at 6AM and return at 5PM) are also sampled in the FHS along the U.S. East Coast, The Southeast Headboat Survey, the NE Vessel Trip Report, the Large Pelagic Survey, or through state programs depending on the region of the United States (Chromy et al., 2009). Anglers on for-hire vessels often may be noncoastal residents, and the random PPS sampling can be a mismatch when the vessels follow strict departure and return schedules. Therefore, their catch and effort might not be well sampled through the APAIS. The FHS was not well sampled in the MRFSS/Coastal Household Telephone Survey, and NMFS developed a separate survey for the Gulf Coast in 2000 and the Atlantic Coast in 2005. West Coast states and Texas retained their programs, and their sampling designs have been reviewed by the MRIP (Chromy et al., 2009).

The sampling frame for the effort component of the FHS is a comprehensive directory of for-hire boats, stratified by vessel type, state, and week. The MRIP website states, "Data collection is conducted on a weekly basis." The contact person for the vessel that has been chosen for the week is mailed a notice and a log sheet. Within a stratum, sampling is done as a systematic random sample without replacement from the stratified list frame (NMFS, 2014b). The respondent can FAX the report, use a toll-free telephone number, or wait to be called by NMFS contractors.

> Respondents are asked to report vessel fishing activity for the prior week, and then asked to profile each for-hire fishing trip. Information obtained for each trip includes area fished, number of anglers who fished, hours of actual fishing activity, method of fishing, and target species, if any....
>
> Effort estimates are produced from the average number of angler-trips per vessel-type per week and the number of vessels per vessel-type in the sampling frame. Adjustment factors for active for-hire fishing boats that are not in the sample frame (new to fleet, no contact information known, etc.) are produced from APAIS questions and applied to the raw effort estimate. (NOAA, 2014, p. 3)

This sampling represents a stratified systematic random sample, which is well understood and has well-known variance properties. The sampling unit is the vessel, and 10 percent of the sampling units are chosen from the frame each week (NMFS, 2014b). CPUE estimates are obtained from interviews of intercepted for-hire trips in the APAIS. Compared to the MRFSS, the improvements to this onsite survey address the concerns detailed in the 2006 NRC report. The onsite survey that obtains CPUE for the FHS is now a probability sample. The MRIP now recognizes that site data should be seen as a cluster, not as a simple random sample of anglers.

The FHS frame for Virginia through Maine includes vessels with highly migratory species (HMS) and large pelagic permits. The survey questionnaire for the FHS queries vessels that caught large pelagics and highly migratory species. A separate biweekly telephone survey is conducted using the FHS frame to estimate large pelagics taken by private boats with HMS permits.

The FHS overlaps with other surveys on the Atlantic Coast, The Northeast Fisheries Science Center's Vessel Trip Report (VTR) program from Maine to Virginia, and the Southeast Regional Headboat Survey, as well as state logbook programs. The MRIP website states, "The VTR data are not used for preliminary wave-by-wave estimates, but they are included at the end of the year when the VTR data are most complete. For all federally-permitted charter boats and headboats, the total trips reported in the VTRs are used to produce an unadjusted number of angler trips. These boats are treated as a separate 'VTR boats' stratum within each for-hire boat mode. All FHS data obtained for those vessels are removed, and FHS estimates of the numbers of angler trips on non-VTR boats are re-run for each wave using the remaining FHS data. The resulting FHS estimates represent a second 'non-VTR boats' stratum for each mode." Initially the VTR reports were evaluated on a yearly basis, but NMFS seeks to obtain these data bimonthly for inclusion in wave-by-wave estimates.

The 2006 NRC report recommended that the for-hire sector be handled as a separate commercial sector. However, the heterogeneity of the for-hire vessel types complicates a singular approach to estimating catch and effort. Boat size, passenger capacity, whether the boat has a regular docking site, species targeted, and permits needed for targeted species vary by region and state. States and the U.S. Coast Guard (for vessels carrying 10 or more passengers) require licensing to operate charter vessels, thus providing a potential list frame for sampling. In addition, charter vessels targeting highly migratory species must have a federal permit to enter the fishery. For-hire vessels that fish for reef fish or pelagics, among other species, must also obtain special federal permits. The advantage in having a for-hire license is the ability to mandate reporting of effort and catch as a provision of for-hire license renewal. However, with a list frame, catch is self-reported and not directly verifiable without onsite validation. Moreover, the value of these data relies on enforcement of the reporting provisions in a timely manner. The level of enforcement varies depending on the regulatory agency and the robustness of its laws.

In a pilot study of electronic logbooks in the Gulf of Mexico for-hire fleet, Donaldson et al. (2013) noted that captains who did not report regularly (or at all) were allowed to file a year-end report and to continue this practice thereafter. Without penalties for this practice, data quality can be compromised because of a lack of validation. Donaldson et al. (2013) recommended that year end–only reporting be penalized, that reporting be done weekly, and that late or missing reports be quickly identified and participants notified quickly of their noncompliance. They also emphasized the importance of field validation, such as by on-board observers depending on vessel size and cooperation, and through onsite intercepts.

Total catch estimation from the FHS mirrors that of the MRIP in that it is a complemented design. Effort is obtained from sampling or censusing a list frame of vessel captains' business telephone numbers. Catch is estimated as CPUE from the APAIS. The proportion of unlisted boats is adjusted based on a ratio derived from APAIS intercepts of angler trips on for-hire vessels that are not on the FHS frame. Because the FHS telephone survey draws from businesses holding state and federal licenses, it should be reliable as a sampling or census frame.

CONCLUSIONS AND RECOMMENDATIONS

Conclusion: The new Access Point Angler Intercept Survey design reflects a substantial improvement of the MRFSS intercept survey methodologies.

Conclusion: The new Access Point Angler Intercept Survey design uses probability proportional-to-size (PPS) sampling at the first stage. This design is expected to lead to efficient estimates if the cluster inclusion probabilities are (approximately) proportional to the cluster catch. However, PPS sampling designs tend to be vulnerable to units whose pressure estimates are poor—high-pressure sites with low catch or low-pressure sites with high catch—and can potentially cause high variance estimates.

Recommendation: The appropriateness of probability proportional-to-size sampling should be evaluated, and alternative sampling designs should be considered if needed. For example, a stratified design (based on the site pressure as a stratification variable) may avoid very small selection probabilities, which, in turn, may lead to more stable estimates. Otherwise, methods dealing with influential values should be considered, including weight smoothing (Beaumont, 2008) and weight trimming procedures (Potter, 1990).

Recommendation: For data users requiring domain estimates at a fine level, design-based estimators tend to exhibit very large variances. To address this, small area estimation procedures should be investigated for obtaining estimates for small domains.

Conclusion: Private sites have zero inclusion probability in the sample. A ratio estimator is used to compensate for the undercoverage. However, the validity of the ratio estimator relies on the assumption that the behavior of private sites is similar to that of public access sites. Otherwise, the ratio estimator may be biased, especially if the proportion of private sites is appreciable.

Conclusion: An additional question on the Fishing Effort Survey about angler use of private or public access sites will enable stratification of the respondents in the Fishing Effort Survey into two strata: (1) anglers who have used private sites and (2) anglers who have used public-access sites. Selecting a sample from each stratum and asking the selected households about their catch will make it possible to assess differences between private- and public-access sites in terms of catch per unit effort. This will provide some insight about the quality of the ratio estimator used to obtain an estimate of *Total Catch*.

Recommendation: NMFS should conduct pilot studies to determine the optimal method for collecting accurate information on total catch differences between public- and private-access sites. For example, NMFS could add a question to the Fishing Effort Survey questionnaire about angler use of private sites or public-access sites. Geographic maps used to identify public-access sites within the state (see Chapter 3) could be used to distinguish public-access from private sites.

Conclusion: Missing data in the Access Point Angler Intercept Survey occur because of refusals (or mid-interview refusals), language barriers, or missed eligible anglers. Missing values may lead to biased estimators if the behavior of nonresponding anglers differs from that of responding anglers.

Recommendation: Interviewers who administer the Access Point Angler Intercept Survey should attempt to collect some paradata to help to reduce the potential bias due to missing interview data.

Recommendation: Anglers are expressing a growing interest in reporting their catches electronically (use of tablets and smartphones). NMFS should conduct a study to compare angler reporting of catch via an app with angler reporting via the traditional interview.

Conclusion: As concluded by the 2006 National Research Council committee, the magnitude and fate of fish discarded by recreational anglers remains highly uncertain. Although some technological changes (e.g., iSnapper) have been incorporated into MRIP data collection, lack of validation of discard estimates significantly contributes to the uncertainty in assessing the impact of discard mortality on stock productivity estimates and management of stock removals.

Conclusion: Initiatives by other branches of NMFS to address discard mortality estimation have not been integrated into MRIP design or operational procedures.

Recommendation: The MRIP should develop and then incorporate validation programs for the estimation of the numbers of fish discarded at sea by recreational anglers. These programs should integrate with other NMFS initiatives concerning estimation of discard mortality.

Conclusion: Recent pilot studies have demonstrated the value of using electronic logbooks to record catch and effort in sectors of the for-hire fisheries. Technological advances have reduced the costs of this equipment, increased the ease of use, and provided value-added benefits to sectors of the for-hire fleet.

Recommendation: The MRIP should expand the electronic logbook program to include most of the large charter and for-hire fleets, through outreach training in electronic logbook use and through implementation of software to run on standard tablets or smartphones.

Conclusion: During the past 10 years, there has been a substantial number of methodological studies/improvements/modifications to both the Fishing Effort Survey and Access Point Angler Intercept Survey. However, the available documentation is not always clear and up to date.

Recommendation: The MRIP should invest resources to provide organized and up-to-date documentation that describes in detail each step of the Fishing Effort Survey and Access Point Angler Intercept Survey methodologies and any changes made to them.

5

Framework for Continued Scientific Evaluation, Review, and Certification

INTRODUCTION

In its 2006 *Review of Recreational Fisheries Survey Methods* the National Research Council (NRC) identified the need to develop an expanded and better structured framework for continued scientific evaluation, review, and certification of the methods, protocols, and procedures used for recreational fisheries data collection. Implicit in this advice was the notion that the Marine Recreational Information Program (MRIP) would need to be flexible and adaptive if it were to become the umbrella framework under which recreational fisheries data for assessment and management at the national level are provided. In other words, an improved program for collecting recreational fisheries data would need to account for the wide-ranging regional differences in U.S. recreational fisheries—and, in many cases, the differences among fisheries within the same region. It would need to develop the capacity to continually assess the strength of the scientific process, including the engagement of external scientific and technical expertise, to develop, test, review, and certify new sampling and estimation procedures. More specifically, the 2006 NRC report recommended that a permanent and independent research group should be established and funded to continuously evaluate the statistical design and adequacy of recreational fishery surveys and to guide necessary modifications or new initiatives. The report went further to say that "human dimensions" expertise should be included as well. This chapter discusses the extent to which these recommendations have been addressed and evaluates whether the framework for continued scientific evaluation, review, and certification established by the MRIP is sufficient and adequate for effective implementation of U.S. marine recreational fisheries surveys.

DEVELOPMENT OF TECHNICAL EXPERTISE

The 2006 committee specifically recommended that a survey office devoted to the management and implementation of marine recreational surveys be developed. This recommendation was based on the recognition that coordination and implementation of marine recreational surveys nationwide is a large and complex task that requires focused and dedicated attention and that expanding staff capacity and developing technical expertise were key to implementing the recommended programmatic improvements. Below the committee discusses the degree to which this recommendation was addressed.

Staffing Improvements

The MRIP has made significant progress in expanding the number of staff dedicated to the program. Since 2006, the number of full-time MRIP staff has increased from 6 to 12, and the program is supported by six full-time contractors. The program has also invested in formal training of existing staff. For example, most of the current MRIP staff have taken graduate-level courses on high-level technical topics such as survey methodology, sampling theory, and survey operations through the Joint Program in Survey Methodology (JPSM).[1] Furthermore, two MRIP staff have completed a master's degree in survey methodology through JPSM. Our discussions with regional and state partners revealed that these staff improvements have greatly increased the MRIP's ability to expand technical support and achieve better regional coordination. In this regard, the MRIP has achieved a desired level of competency and, in conjunction with consulting of external experts, has made excellent progress since 2006.

The MRIP Communication and Education Team (CET) could still benefit from additional staffing. This team is currently supported by only one full-time and two part-time contractors and is tasked with the development, implementation, and coordination of the MRIP's communications strategy nationwide. The success of the MRIP depends to a large degree on clear, accurate, and timely communications and on engagement of all stakeholder groups, including anglers. Therefore, the MRIP should consider expanding its communications staff to better address this important task. Chapter 7 provides a more complete review and discussion of the MRIP's communications and outreach program.

[1]Founded in 1993, JPSM is the nation's oldest and largest program offering graduate training in the principles and practices of survey research. JPSM is sponsored by the Federal Interagency Consortium on Statistical Policy and located at the University of Maryland. To date, it has more than 240 graduates working in government agencies, academic settings, and private survey research firms.

Workshops, Conferences, and Symposia

The MRIP has either organized or been closely involved in the organization of several recreational fisheries workshops or symposia conducted either in coordination with regional partners (e.g., the regional Fisheries Information Networks [FINs] and the Atlantic Coast Cooperative Statistics Program [ACCSP]) or conducted as part of national or international scientific conferences (Table 5.1). These workshops and symposia have been attended by highly trained specialists and experts in the field and, therefore, facilitated review and discussion of MRIP issues by a broad range of scientists, promoted cross-pollination and exchange of ideas, and exposed MRIP technical staff, as well as regional and state partners, to a variety of recreational fisheries issues under different scenarios.

Consultants

The 2006 NRC report recommended that a permanent and independent research group should be established and funded to continuously evaluate the statistical design and adequacy of recreational fishery surveys and to guide necessary modifications or new initiatives. In response, the MRIP has developed and maintained a high-end cadre of statistical consultants who have greatly advanced survey revisions and improvements, facilitated faster and broader implementation of the MRIP certification process, and increased the efficiency of providing technical advice and guidance to regional and state partners.

An issue still to be addressed is the need for additional consultant support. The committee received feedback during its regional meetings that indicated that, despite the high quality of the current consultants' input and advice, the review and certification process remains relatively slow, mainly because many regional and state partners have requested advice on survey improvements or submitted documentation for certification of new surveys. At the same time, this same pool of consultants assists the MRIP in other capacities, for example, by creating and testing new survey methods and improving existing survey designs. Expanding the pool of experts serving as MRIP consultants would allow for a faster, more efficient review process, as well as ensure a continuous infusion of energy and ideas.

DEVELOPING, TESTING, AND IMPLEMENTING NEW TECHNIQUES

In 2008 the MRIP established a Pilot Studies Program for developing, testing, reviewing, and eventually certifying new sampling and estimation procedures to be applied under the MRIP umbrella. Most of these pilot studies have been developed and implemented in collaboration with state and regional partners, usually by efforts coordinated through the regional Interstate Marine Fisheries Commissions and their associated FINs (see Box 5.1).

The MRIP Pilot Studies Program is implemented in three concurrent phases: (1) evaluation of current methods, (2) innovation to identify and test new meth-

TABLE 5.1 List of Workshops and Symposia Organized by the MRIP on Recreational Fisheries Issues

Name of Event	Date
MRIP workshop: Recreational Fisheries Statistics Requirements Management Framework Workshop	September 2006
MRIP workshop: Marine Recreational Information Initiative Operations Team Workshop	August 2007
International Council for the Exploration of the Seas (ICES) Annual Science Conference Session on Survey Methods for Recreational and Artisanal Fisheries	September 2008
MRIP workshop: Review of For-Hire Recreational Fisheries Surveys	March 2008
ICES Workshop on Survey Methods for Recreational Fisheries	June 2009
New Zealand Workshop on Recreational Fisheries Survey Methods	August 2009
Workshop of ICES Planning Group for Recreational Fisheries Surveys (PGRFS)	June 2010
Norway Workshop on Recreational Fisheries Survey Methods	June 2010
MRIP workshop: Review of Oregon Recreational Fisheries Surveys	July 2010
MRIP workshop: Review of Washington Recreational Fisheries Surveys	November 2010
MRIP workshop: Addressing the Fishery Management Need for More Timely Recreational Data	March 2011
Workshop of ICES Planning Group for Recreational Fisheries Surveys (PGRFS)	April 2011
MRIP workshop: Review of California Recreational Fisheries Surveys	June 2011
6th World Recreational Fisheries Conference Session on "New Methodological Tools to Survey and Assess Recreational Fisheries"	August 2011
American Fisheries Society (AFS) Symposium: Improving Survey Methods for Monitoring Recreational Fishing Effort and Catch	September 2011
MRIP workshop: Volunteer Angler Survey Workshop-Inventory Existing Programs and Assess Utility of volunteer angler surveys	February 2012
MRIP workshop: MRFSS/MRIP Calibration Workshop	March 2012
Workshop of ICES Working Group for Recreational Fisheries Surveys (WGRFS)	April 2012
MRIP workshop: A Review of the Current Sampling and Estimation Methods of the Hawaii Marine Recreational Fishing Survey (HMRFS)	July 2012

TABLE 5.1 Continued

Name of Event	Date
MRIP workshop: Evaluation and Improvement of the Current Sampling and Estimation Methods for the Puerto Rico Recreational Fishing Survey	September 2012
MRIP workshop: Development of a Survey Design(s) for Collecting Recreational Fishing Data in the US Virgin Islands	September 2012
MRIP workshop: MRIP Methods and Data Seminar with NY Stakeholders	April 2013
Workshop of ICES Working Group for Recreational Fisheries Surveys (WGRFS)	April 2013
MRIP workshop: Design Effort Surveys for Shoreline Fishing in Highly Migratory Recreational Fisheries Surveys	July 2013
MRIP workshop: Marine Recreational Information Program Executive Steering Committee Implementation Workshop	July 2013
ICES Annual Science Conference Session on Marine Recreational Fisheries: Understanding Impacts and Consequences for Management	September 2013
MRIP workshop: Gulf of Mexico Red Snapper Recreational Catch Accounting Methods Workshop	November 2013
MRIP workshop: Gulf of Mexico Red Snapper Recreational Catch Accounting Methods Workshop II	March 2014
Workshop of ICES Working Group for Recreational Fisheries Surveys (WGRFS)	June 2014
MRIP workshop: Proportional Standard Error and Management Uncertainty in Recreational Data Collection on the Atlantic Coast	September 2014
MRIP workshop: MRIP Calibration Workshop II	September 2014
MRIP workshop: Gulf of Mexico Red Snapper Recreational Catch Accounting Methods Workshop III	December 2014
MRIP workshop: Peer Review of Louisiana Creel Survey Program	June 2015
AFS Symposium: Survey Methods for Monitoring Recreational Fisheries in Support of Stock Assessments and Fisheries Management	August 2015
MRIP workshop: Peer Review of Alabama Snapper Check Survey Program	December 2015
MRIP workshop: For-Hire Programs: Inventory, Certification, and Integration Planning	May 2016
MRIP workshop: Peer Review of Mississippi Tails 'n Scales Survey Program	June 2016

SOURCE: Committee.

BOX 5.1
Regional Interstate Marine Fisheries Commissions and the Fisheries Information Networks

In the 1940s, the federal government authorized by statute the establishment of three interstate compacts, each creating a regional marine fisheries commission to better utilize and protect fisheries within the consenting states' jurisdiction. The three separate commissions represent the Atlantic (Atlantic States Marine Fisheries Commission), the Gulf of Mexico (Gulf States Marine Fisheries Commission), and the Pacific states (Pacific States Marine Fisheries Commission).

The regional Fisheries Information Networks (FINs) and the Atlantic Coast Cooperative Statistics Program (ACCSP) are state-federal cooperative programs in which NMFS participates as a partner with the state fisheries agencies, interstate marine fisheries commissions, regional fishery management councils, and other federal agencies, such as the U.S. Fish and Wildlife Service. These partnerships engage in cooperative programs to collect, aggregate, and manage state and federal fisheries data (both commercial and recreational) to support fishery managers and associated agencies. The regional FINs and ACCSP also serve as liaisons for identifying state and regional data needs and are, therefore, well positioned for coordinating the funding and implementation of regional MRIP pilot studies.

ods, and (3) implementation of proven methodologies. Funding—about $14.7 million from 2008 to 2015—has been largely provided by the National Marine Fisheries Service (NMFS) through a grants program coordinated through the regional FINs and ACCSP. The MRIP Operations Team solicits and reviews research proposals and provides recommendations for funding based on resources available (Table 5.2).

Table 5.3 provides examples of recently funded MRIP pilot projects and the various data or methodological needs they have addressed. A total of 105 regionally based pilot projects have been conducted since the program was implemented in 2008 in response to the NRC 2006 review.

Our review of the MRIP pilot study program indicates that it constitutes an appropriate and effective mechanism for providing highly specialized technical and scientific support (including access to technical consultants) for the development, review, and certification of regional- or state-specific surveys. This increased capability has greatly enhanced the MRIP's capability to address regional and state needs for stock assessment and fisheries management.

TABLE 5.2 Summary of the Processes Led by or Functions Provided by Different Organizations, Institutions, and Entities Involved in the Development and Implementation of MRIP Pilot Projects

Institution or Organization	Function or Process
MRIP Team (Executive Steering Committee, NMFS Office of Science and Technology)	• Develop and conduct pilot studies • Manage peer reviews • Identify and recommend best practices • Facilitate regional implementation • Manage implementation (certain regions)
Regional Partners (States, Interstate Fisheries Commissions, RecFINs, Councils, NMFS Regional Offices and Science Centers)	• Identify region-specific data needs and priorities • Adapt certified methods to meet regional needs, and, as necessary, secure additional resources • Manage implementation
Stakeholders (Anglers, Charter boat owners/operators, Nongovernmental organizations)	• Work with MRIP team to identify research needs and, as appropriate, lead or participate in pilot studies • Work with regional partners to identify data needs and priorities, and, as appropriate, assist with resource acquisition

SOURCE: Committee.

TABLE 5.3 Examples of Recent Pilot Projects Funded by the MRIP

Pilot Project Title	Summary Description
Alaska For-Hire Electronic Logbook Census	Implementing new electronic logbook program targeting fishing guides and guide businesses in Alaska.
Hawaii On-site Private Boat Catch Survey	Building on work that has taken place in the Gulf of Mexico and Atlantic Coasts, design and test an appropriate onsite survey to estimate catch rates from private fishing boats in Hawaii.
North Carolina and South Carolina For-Hire Electronic Logbooks	Developing NC and SC for-hire industry logbook reporting programs, with supporting validation survey designs that can be used to meet any existing federal logbook reporting requirements.
Alabama Private Boat Electronic Red Snapper Fishing Census	Refining the reporting methods and field validation protocols for reporting recreational red snapper landings by Alabama private recreational vessels.
Testing the Impacts of One-Month Waves	Assessing the potential for bias resulting from measurement error in the Fishing Effort Survey and evaluating the impact of 1-month versus 2-month reference waves on the precision and timeliness of estimates on the Gulf of Mexico and Atlantic Coasts.

SOURCE: Committee.

USE OF NEW TECHNOLOGIES

Traditionally, recreational fishing surveys have been conducted by trained interviewers who collect specific data about anglers' fishing trips and the nature of the species they landed and released on paper-and-pencil survey forms. However, there has recently been a growing interest throughout the recreational fishing community in identifying scientifically sound, statistically robust methods to report recreational fisheries data electronically (e.g., via smartphones and tablets). The use of these new technologies could potentially improve the timeliness and accuracy of recreational fisheries data, as well as reduce costs and paperwork burdens (e.g., nightly submissions of data to increase the timeliness of recreational catch and effort estimates to the public).

The MRIP has been evaluating and testing the use of these new technologies primarily through four distinct initiatives:

1. **For-hire electronic logbooks**. The MRIP indicated that it has developed a comprehensive road map for implementing electronic reporting in the for-hire sector. It incorporates the work of three separate pilot studies and identifies the minimum requirements for ensuring success. These requirements include compliance measures, the need for hardware and software development standards, and the need for statistically sound designs for combining electronic logbook data with data collected in independent dockside and/or at-sea surveys for the purposes of validation.
2. **Angler electronic data reporting**. With increases in smartphone use and Internet access, intense interest in the use of electronic reporting technologies by individual anglers has arisen. The MRIP seems engaged in further developing and expanding the electronic reporting to provide angler-provided catch data that are usable and statistically valid, exploring options that can be incorporated into, as well as supplement, existing surveys. Building on previous work in this area, the MRIP is funding several separate pilot studies to examine electronic reporting options to allow anglers to self-report data electronically.
3. **Sampler electronic data capture and submission**. The MRIP has also been testing the use of tablets and other electronic data-collection platforms to allow dockside samplers to capture and submit data electronically. Such platforms could accelerate the provision of more timely data, reduce or eliminate the potential for data transcription errors, and facilitate implementation of real-time checks of data ranges and corrections at the data-collection stage (i.e., automated quality assurance and quality control [QA/QC]).
4. **Use of electronic monitoring for validation purposes**. As noted previously, the validation of self-reported data on discards of recreationally caught fish at sea is important to the provision of accurate estimates of total removals by the recreational fishery. While observers can provide

> both validation and biological sampling of discarded fish, the recreational fishery is not well suited to deployment of observers. The use of cameras to validate total catch is becoming more common in commercial fisheries (Wallace et al., 2015) and may improve estimates of discards within recreational fisheries. Pilot studies have deployed electronic monitoring systems on even small charter vessels (< 30 ft) with good success. The committee believes that the MRIP could benefit from increased consideration and application of electronic monitoring for some sectors of recreational fishing.

In practice, evaluation and testing of new technologies for MRIP fisheries data collection are being accomplished through implementation of several MRIP-funded pilot studies, often structured according to Regional MRIP Implementation Plans. For example, researchers and stakeholders in Florida are testing the use of a smartphone- and Internet-based electronic reporting tool called iAngler to collect and report data on recreational effort and catch. A similar project is being implemented in Texas to test the use of an electronic reporting tool called iSnapper for collection of self-reported recreational fisheries catch data.

Despite these efforts, stakeholders indicated to the committee that large portions of the private angler and for-hire sectors perceive that implementation of electronic reporting or adoption of smartphone applications for volunteer angler self-reporting is not happening fast enough. This seems to be largely a communications issue—that is, anglers and for-hire captains are acutely aware of the potential for developments in smartphone, tablet, and other portable technologies, and see the potential to optimize data-collection methods. The MRIP and its partners see the advantages of electronic reporting but, given their technical roles, consider this to be primarily a statistical issue—that is, they are being cautious because they do not want to compromise the statistical robustness of survey estimates in the name of increased sampling efficiency and participation. Thus, the MRIP's implementation of electronic reporting has been mostly exploratory and focused on testing different formats and platforms, and developing statistical techniques for integrating electronic-based data-collection programs into the existing survey estimation procedures (see Table 5.3 above for some examples).

Input from regional partners, anglers, and for-hire operators suggests that the delay in implementing electronic platforms for collection of recreational fisheries data is perceived differently by different groups. Fisheries scientists and managers as well as regional partners (i.e., regional coordinating staff associated with the RecFINs and ACCSP) are more understanding of the MRIP's decision to delay full implementation of electronic reporting until scientifically valid estimation techniques are fully developed and properly tested. A large component of the private angler and for-hire sectors feels differently, probably because the integration of electronically reported data into the standard MRIP estimation procedures is highly technical and statistically complex and far from intuitive. Thus, there again appears to be an issue for the communications team.

Other challenges exist. The flexible and distributed MRIP model—that is, surveys are developed and conducted according to regional implementation plans and with close coordination and input from state partners—makes adoption of electronic reporting and electronic platforms for data capture by MRIP samplers (i.e., instead of pencil and paper) more complicated. Issues such as the willingness of state partners to comply with or accept the use of electronic platforms for data capture and the costs involved in implementing these platforms nationally still exist.

In general, given the many complexities and challenges involved, the MRIP has done a good job of evaluating the use of new and emerging technologies for electronic reporting of recreational fisheries data. However, the perception by many that the MRIP is moving too slowly in incorporating these technologies needs to be addressed. The committee also sees potential for gains in accuracy through electronic monitoring of discards. The MRIP should develop a strategy to better articulate the complexities, costs, and timelines associated with the implementation of new and emerging technologies in recreational fisheries data collection. This communication strategy should not only focus on regional and state partners but also address the questions and concerns expressed by private anglers and for-hire operators (see Chapter 7 of this report for a broader discussion of the MRIP's communication and outreach).

THE MRIP CERTIFICATION PROCESS

The 2006 NRC report recommended that a permanent and independent research group should be established and funded to continuously evaluate the statistical design and adequacy of recreational fishery surveys and to guide necessary modifications of new initiatives. In response to this recommendation, the MRIP has established a rigorous and systematic peer-review process to ensure that new survey and estimation methods are scientifically sound before they are "certified" and made available for use in stock assessments and fisheries management. Furthermore, the MRIP certification process incorporates detailed QA/QC requirements to reduce errors in the estimates produced by these add-on or supplementary surveys. Once certified, methods are available for use by the MRIP and its partners.

The process for MRIP certification seems fair and appropriate. In general, the MRIP only supports projects (i.e., financial and logistical support, including access to technical staff and statistical consultants) that apply methods that have been MRIP certified. The MRIP may support the use of methods that have not been certified if a plan to certify those survey methods is in place and being followed. The certification process is focused on evaluating new or replacement survey and estimation methods, and modifications or recommended improvements to existing methods. To be granted final MRIP certification, survey methods must

- Adhere to applicable MRIP standards and best practices (specific documentation for these standards are available at the MRIP website),
- Be peer reviewed and supported by the results of the review, and
- Be approved by the MRIP Operations Team, MRIP Executive Steering Committee, and NMFS leadership.

Our review of the criteria and procedural steps involved in the MRIP certification process (Table 5.2), as well as input from and discussions with multiple MRIP regional and state partners, indicates that, in general, the process is working well. For example, in 2010 the MRIP funded a full review of the Oregon Department of Fish and Wildlife Marine Resources Program's Ocean Recreational Boat Survey and the Washington Department of Fish and Wildlife's Ocean Sampling Program and Puget Sound Sampling Program, and in 2011 the MRIP funded a review of the California Recreational Survey Program. The committee received feedback that the partners greatly appreciated the input and reviews. The process was very interactive, involved multiple consultations with MRIP staff and statistical consultants, and provided an opportunity for those state agencies to make significant progress on developing and testing potential survey improvements. In early 2015, both the Oregon Department of Fish and Wildlife Marine Resources Program and the Washington Department of Fish and Wildlife submitted materials detailing their survey designs and programs for MRIP review and certification. Hawaii and Louisiana are undertaking similar processes.

The only major concerns the committee heard regarding MRIP certification related to the length of time associated with the process (i.e., perceived delays, lack of timeliness) and the uncertainty regarding the level of funding required to implement review recommendations. For example, the base level of survey funding for Pacific Coast states has been flat for many years, representing less than 50 percent of the overall costs of the marine recreational angling surveys. Flat or reduced funding has made implementation of recommended survey improvements difficult and in some cases impossible for states to accomplish.

The lack of timeliness in the MRIP certification process seems to be related to a need for additional people to conduct technical reviews or serve as statistical consultants (e.g., statisticians, survey methodologists, data-collection experts, human dimension and cognitive scientists). As noted above (see "Development of Technical Expertise"), the significant expansion of MRIP staff since 2006 has greatly benefited the program and has been highly praised by the MRIP's regional and state partners. In addition, engaging statistical consultants with high levels of expertise shown by the group of statistical consultants the MRIP has engaged in the review and certification process is highly commendable. The problem seems to be rooted in the need to expand the number of people involved in this process so that review and certification of multiple survey programs in different regions can occur simultaneously and consistently. Because most programs are implemented in part through funding/governance by state Marine Fisheries Commissions (e.g., PSMFC), any alterations required by certification review would need to be coordinated with such agencies, which presents a strong argument for the simultaneous certification of state programs.

The review committee also discussed whether mechanisms are in place to confirm that surveys are implemented according to certified protocols and that assessments of data quality occur before incorporation of state or regional survey data into the MRIP database. Fortunately, because the vast majority of surveys certified by the MRIP are implemented through Regional Implementation Plans, several mechanisms are in place (e.g., regional RecFIN meetings, MRIP Wave meetings, regional stock assessment panel meetings) to ensure that approved survey protocols and collected data meet quality standards.

TRAINING OF INTERVIEWERS

Chapter 3 discussed interviewer effects and interviewer variation and their contributions to nonsampling error. One way to minimize errors due to interviewer variation is to provide appropriate and rigorous interviewer training (Dahlhamer et al., 2010). Standardized training will also facilitate the comparison of data at the state, regional, and national levels. Several MRIP documents reflect concern on the part of the program to produce well-trained interviewers. These include an outline of training structure, coordination and quality control (NOAA, 2016b), and procedures for interviewer training (Procedures Manual: Access Point Angler Intercept Survey [APAIS], January 2014). These documents reflect a genuine concern for maintaining training standards and highlight the following areas: interviewer knowledge (e.g., fish identification), concept definitions, basic sample design, survey data-collection procedures, survey structure, conduct of the creel survey, administrative issues, and materials and supplies. Although there is some practical, albeit limited, information on dealing with, for example, possible respondent hostility onsite, the procedures manual and statement of work (SOW) focus primarily on the technical aspects of conducting the APAIS. The following topics have not been directly addressed in documents and require further consideration.

Not everyone is well suited to be an interviewer: Recruitment of competent interviewers is critically important and should be addressed from a human resource perspective.

Improve quality control through statistical assessment: The documents, particularly the SOW, discuss aspects of QC in relation to interviewer training, evaluation, and monitoring. They mostly concern QC with regards to interviewer knowledge (e.g., fish identification), onsite monitoring, telephone verification of intercept surveys, and data quality checking. However, with respect to data QC, the documents focus on the use of error-checking software, which generally seeks and flags outlier entries for further scrutiny. Although these methods can improve data quality by detecting recording errors and data entry errors, it does little to detect and address errors due to the interviewers themselves. As discussed in Chapter 3, statistical methods that can detect and model interviewer error should be considered as a supplemental method for increasing data quality through the QC efforts.

Not all training is purely of a technical nature: Although the training materials are technically sound, they convey a limited amount of practical experiential knowledge. In part, interviewing is an art and thus necessitates training materials that can provide more qualitative insights and guidelines for interviewers in the training process. It might be useful to produce an "interviewing in the MRIP context handbook" that documents interviewers' experiences and methods for dealing with the vagaries of interviewing in the field. This would constitute an interviewer's field guide with information on potential problems and solutions for interviewing in the APAIS context. The committee judges that development of this handbook would greatly benefit from input from the MRIP CET.

CONCLUSIONS AND RECOMMENDATIONS

Conclusion: During the pilot study phase of the Fishing Effort Survey (FES), NMFS has benefited from access to an independent group of statisticians and survey methodologists with substantive experience. The ever-changing world of survey research will require continuous evaluation of the FES, including experimentation, to ensure high-quality estimates in a timely manner.

Conclusion: If NMFS could expand the existing capacity of this consultant pool, both in number and in expertise (e.g., experts in cognitive issues, including angler recall), duplication of effort would be reduced and the provision of technical and scientific support would be facilitated. In addition, the MRIP certification process would be streamlined. Any such group should be periodically refreshed to include new researchers with a variety of interests and expertise.

Conclusion: With the development of a certification process, the MRIP made substantial progress toward implementing relevant key recommendations of the 2006 National Research Council report. The MRIP has invested in the development of a well-structured process for continued scientific evaluation, review, and certification of the recreational fisheries surveys conducted under its umbrella. This process also affords a mechanism for providing highly specialized technical and scientific support (including access to technical consultants) toward the development, review, and certification of regional- or state-specific surveys.

Conclusion: The MRIP has made progress in evaluating and testing the use of new technologies (i.e., smartphones, tablets, and other electronic data-capture platforms) as ways to implement electronic reporting, avoid or decrease data transcription errors, and increase the timeliness and reliability of recreational fisheries data collection. Still, the impression among many private anglers and the for-hire sector is that implementation of these technologies is not occurring quickly enough.

Recommendation: The MRIP should develop a strategy to better articulate the complexities, costs, and timelines associated with implementing new and emerging technologies in recreational fisheries data collection and monitoring. This communication strategy should not only focus on regional partners but also address questions and concerns expressed by private anglers and for-hire operators. It should involve both the MRIP communications team and the NMFS Office of Communications.

6

Degree of Coordination

INTRODUCTION

The multijurisdictional nature of marine fisheries management, which in most regions of the country involves not only the regional fisheries management councils but also multiple states and institutions, presents myriad coordination challenges to data collection, data management, stock assessment, and ultimately fisheries management. To collect recreational fisheries data that meet required standards for assessment and management in this complex, multijurisdictional system, Marine Recreational Information Program (MRIP) surveys are conducted in cooperation with a variety of regional and state agencies and with the assistance of other institutional partners. The National Research Council's (NRC's) 2006 report recommended that a greater degree of standardization among state surveys, and between state surveys and the central Marine Recreational Fisheries Statistics Survey (MRFSS), should be achieved. Implicit in this recommendation is the need for a greater degree of cooperation and coordination among the managers of the various surveys. This chapter evaluates whether the degree of coordination among federal, state, and territorial survey programs conducted under the MRIP umbrella is sufficient to support implementation of survey methodologies that address the diversity of regional and state needs while maintaining a clear, cohesive perspective on the nation's marine recreational fisheries (Task 4 of the committee's statement of task).

UNIQUE NEEDS AT REGIONAL AND STATE LEVELS

U.S. marine recreational fisheries show wide-ranging differences across regions and often within regions. These differences can be attributed to interregional variability of several factors, including the length and shape of the coastline and other ocean features, biogeographic patterns in species diversity, and socioeconomic and demographic factors. For example, the South Atlantic and Gulf of Mexico regions have high species diversity, as well as broad, shallow continental shelves and several estuaries, which help to explain their dominance in the magnitude of marine recreational fishing trips nationwide.

The relative importance of individual recreational fishing modes (i.e., private/rental boat, shore-based, and charter/for-hire) also varies widely among U.S. regions. Thus, some regions and states can have survey needs that are specific to stocks. For example, in the Gulf of Mexico, red snapper have relatively short recreational fishing seasons, which present significant challenges to the MRIP, both in estimating catch and effort as well as in monitoring landings. Other stocks may represent "rare-event species" that occur sporadically in the catch and are not properly estimated by the standard MRIP survey approach. In some cases, the proportions of fishing modes may vary significantly across a region. For example, areas with marked differences in the proportion of private anglers versus the charter/for-hire sector require the implementation of more customized MRIP sampling draws so that data collection across the region can reflect these intra-regional differences. Accommodating these regional differences requires the MRIP to adopt a regional implementation approach that is flexible enough to address these unique regional and state needs while maintaining the standardization and national-level cohesion recommended by the 2006 NRC report.

HAS THE MRIP BEEN ADDRESSING REGIONAL AND STATE NEEDS?

Despite the lingering public perception of a centralized, top-down implementation approach the MRIP has, by and large, been responsive to regional and state needs. The 2015-2016 MRIP Implementation Plan Update[1] describes the program as a collaborative, multi-institutional effort focused on developing and implementing a system of surveys that provides the best possible scientific information for use in the management of the nation's marine recreational fisheries. The plan also states that, given the dynamic nature of fisheries and fisheries management practices, the MRIP must be

- Flexible enough to be updated, modified, expanded, or contracted to meet specific regional or local informational needs;
- Robust enough to provide the most precise and least biased information possible;

[1] See http://www.st.nmfs.noaa.gov/Assets/recreational/pdf/FINAL-updated-implementation-plan-3.22.16.pdf.

- National in scope but regionally specific, recognizing that each region has unique informational needs and data-collection issues; and
- Inclusive and transparent, providing scientists, managers, and stakeholders an opportunity to participate in its development and use.

Our review of the MRIP Implementation Plan, as well as input from and discussions with multiple MRIP regional and state partners, indicate the program has made robust progress in achieving these goals. In particular, the MRIP has made great progress in expanding and strengthening the coordination and provision of logistical and technical support to state partners through regional Interstate Marine Fisheries Commissions and their associated Fisheries Information Networks (FINs), as well as the Atlantic Coast Cooperative Statistics Program (ACCSP). Development, coordination, and implementation of regional- and state-specific recreational fisheries surveys conducted under the MRIP umbrella have been largely accomplished through the regional FINs and the ACCSP, often through the establishment of MRIP Regional Implementation Teams. Each Regional Implementation Team is responsible for identifying regional needs and developing a plan to implement improved data-collection designs that address both regional and national needs. Regional MRIP Implementation Plans, which are reviewed and approved by the MRIP Operations Team and the Executive Steering Committee, provide estimated implementation costs and attempt to reach consensus among regional partners with respect to regional needs and implementation priorities.

The growth in the number of state fish and wildlife agencies that conduct the APAIS survey as contractors under MRIP protocols has enhanced coordination between the MRIP and the states as well as expanded opportunities to adjust the survey to address specific regional and state needs. For example, during the past several years, the MRIP and the Gulf States Marine Fisheries Commission, working closely with the five Gulf of Mexico states, conducted a series of workshops that led to the development and implementation of coordinated pilot studies to evaluate several survey methodologies and approaches for estimating catch and effort for Gulf red snapper, a fishery characterized by short federal fishing seasons (e.g., in 2016 the season was 9 days for private anglers and 46 days for the charter/for-hire sector) and unlikely to be properly sampled by the standard MRIP survey protocols. Likewise, territorial governments conduct recreational fishing surveys in the Western Pacific Territories with support from the Western Pacific Fisheries Information Network and the National Marine Fisheries Service (NMFS) Pacific Islands Fisheries Science Center. In Hawaii and Puerto Rico, the MRIP has coordinated with the state fish and wildlife agencies to develop an enhanced survey design to meet the unique needs of the Caribbean Region and island fisheries.

Challenges remain, however. Some state needs—e.g., development and implementation of recreational fisheries catch and effort estimates at small spatial scales to assess or manage state-managed species, or in-season monitoring of compliance with Annual Catch Limits (ACLs)—have been difficult to address. This is particularly true when these needs require a disproportionate increase in sampling effort and become cost prohibitive, or are so specialized (i.e., focused on addressing nontraditional, very specific needs) that their integration into the standard MRIP may affect general survey implementation or compromise the estimation process. For example, fish and wildlife agencies in Alaska and Texas administer marine recreational fishing surveys outside of the MRIP framework, because they judge (1) that the MRIP survey—or family of surveys—cannot provide the estimates of recreational fisheries catch and effort that are needed for assessment and management at smaller temporal and spatial scales; (2) the MRIP surveys cannot address some unique, highly specialized fisheries; or (3) the existing pre-MRFSS/MRIP surveys already in place were sufficient to meet their data needs and implementation of a completely new survey protocol was unnecessary. The Louisiana Department of Wildlife and Fisheries has also been implementing an independent saltwater recreational fisheries survey called LA Creel[2] since 2013. However, they have coordinated with MRIP consultants since then and have applied for MRIP certification, with the objective of ensuring that LA Creel data are compatible with the MRIP and other regional data for stock assessment and management purposes.

However, the committee highlights the importance for alternative surveys, both under the auspices of the MRIP and those of individual states, to be statistically sound. It became apparent during some public testimony to the committee and the committee's internal review that surveys in some areas are not designed to provide estimates with either the precision assumed or the unbiased nature presumed. As such, management of broad-ranging species may be compromised.

Programs in different parts of the country are in different stages of evolution. For example, the MRIP has been working with the Pacific Coast states (Washington, Oregon, and California) to evaluate and test modified survey methodologies that can lead to MRIP certification. However, continued coordination, technical support, and integration of Pacific Coast state surveys into the MRIP framework are warranted. Despite noticeable progress on regional coordination, Pacific Coast states still perceive the MRIP as only a partial solution to their long-term recreational data-collection needs. Furthermore, the base funding from the MRIP to the Pacific Coast states has been flat for many years, representing less than 50 percent of the overall costs of their marine recreational angling surveys. Flat or reduced funding has made implementation of recommended survey improvements difficult and in some cases impossible for the states to accomplish. If this problem persists, any meaningful gains to the states' sampling program through the MRIP certification process are potentially at risk.

[2] See http://www.wlf.louisiana.gov/lacreel.

Finally, it is important to keep in mind that the MRIP was developed primarily to provide estimates of recreational fishing catch and effort for fisheries stock assessment and management at the regional scale (i.e., at the Regional Fisheries Management Council and Interstate Marine Fisheries Commission jurisdictional scales). Development of estimates at smaller geographic scales, although in many cases possible and warranted, might require additional investments at the state or local levels. It is our perception that the increased provision of logistical and technical support by the MRIP, as well as the MRIP certification process, facilitate better coordination and integration of new, more specialized surveys into the MRIP survey framework and that regional and state partners are, by and large, satisfied with the level of support and coordination provided.

OTHER SURVEY PROGRAMS

The NMFS, states, and territorial survey programs conduct several more specialized surveys that fall outside of the MRIP umbrella but require close coordination with the MRIP. This is particularly true when these other surveys (1) address specific fisheries species not likely to be well sampled by the MRIP, (2) are implemented in states where the MRIP does not operate, or (3) are critical complementary components to the MRIP because they focus on specific recreational fisheries sectors. The sections below provide a brief description of these surveys and a discussion of their coordination with the MRIP aimed at providing a clear, national perspective on marine recreational fisheries.

Large Pelagics Survey

Large pelagic and highly migratory species (HMS) such as tunas, billfishes, and some sharks present a special challenge for recreational fisheries surveys. Many of these species are part of "rare event" or "pulse" fisheries—that is, they are caught on a small proportion of all fishing trips and their activity often happens in bursts, as opposed to over a longer season. This necessitates the use of a separate, dedicated survey that both focuses on the characteristics of large pelagic and HMS fisheries and coordinates with the MRIP. On the Atlantic Coast from Maine to Virginia, NMFS uses the Large Pelagics Survey (LPS) to measure the total recreational catch of these species. The LPS includes two complementary survey components. The Large Pelagics Intercept Survey interviews randomly selected anglers and for-hire captains returning from fishing trips targeting large pelagic fishes and measures average catch per trip, average size of kept fish, and number of fish released alive. The Large Pelagics Telephone Survey interviews randomly selected recreational anglers and for-hire captains who hold permits to fish for HMS. It produces the estimates of fishing effort, or the total number of trips taken for large pelagic species during a given time period. Additional biological information is gathered through the Large Pelagics Biological Survey.

This supplemental dockside survey is used primarily for recreationally landed bluefin tuna, targeting both private and for-hire boats. The survey collects length, weight, and body part samples for use by scientists in studies of fish populations and stock assessments. NMFS administers all of these surveys, with input from the HMS Advisory Panel to the NMFS Office of Sustainable Fisheries and under close coordination and oversight by the MRIP.

Southeast Headboat Survey

The Southeast Headboat Survey is a logbook and port sampling program designed and operated by NMFS, Southeast Fisheries Science Center, Beaufort Laboratory, North Carolina. It includes monthly logbook submission of trip-level reports of marine recreational fishing on headboats that target reef fishes and a dockside biological data-collection program for targeted species. The MRIP has supported several pilot projects to improve the documentation and estimation for this program, and it is currently coordinating a pilot test for an electronic data-capture platform for headboat logbook data submission.

Alaska

Four programs funded and fielded by the Alaska Department of Fish and Game (ADF&G), Division of Sport Fish, provide the recreational fishing catch and effort data, and biological data (age, size, and sex composition), necessary to support the North Pacific Fishery Management Council (NPFMC) and NMFS (Alaska Fisheries Science Center, Alaska Region) for federal and international management, primarily of halibut and groundfish. All management of recreational salmon fishing in Alaska is delegated to the state of Alaska through the NPFMC and Pacific Salmon Treaty authorities and regulatory processes.

The U.S.-Canada International Pacific Halibut Commission (IPHC) is responsible for conservation of halibut. Harvest biomass and biological characteristics of Pacific halibut by the recreational sector in Alaska and the Pacific Coast are used by the Pacific Fishery Management Council, NPFMC, and IPHC to assess the coast-wide abundance of Pacific halibut and to allocate Pacific halibut harvests between the recreational for-hire and commercial sectors in IPHC areas 2C and 3A in Alaska and to all users on the Pacific Coast. Independent anglers in Alaska are subject to less restrictive individual angler catch measures, as well as the lack of total catch restriction. The catch and biological data are transmitted to the IPHC each October in the form of a memo that is incorporated into the halibut stock assessment and the Fishery Removals section of the Report of Assessment and Research Activity, as well as to the NPFMC in the form of an oral report and accompanying tables.

Data on harvest biomass and release mortality biomass of demersal shelf rockfish (DSRs) reported by the recreational fishery in the Outside District of southeast Alaska are integrated into the stock assessment of DSRs in this area. These data are transmitted via email to the ADF&G Commercial Fisheries Division each October for development of the Stock Assessment and Fishery Evaluation report for this stock.

The ADF&G salmon surveys not only provide salmon catch, effort, and biological data, but also produce data on recovery of coded-wire tags for estimating hatchery and wild stock contributions of salmon.

Alaska has not traditionally participated in MRFSS or MRIP activities. In large measure, this is because of a perception that management of resources in this area was restricted to concerns solely within the state and because the methodology used to estimate recreational catch has been well developed for some time (see Mills, 1979; Mills et al., 1986). The methodology used is similar to that of the MRIP—a mail survey to estimate fishing effort and an intercept survey for catch per unit effort. Enhancements over time have included mandatory charter logbook regulations and electronic reporting. In this sense, the Alaska program anticipated the methodology currently used by the MRIP. The committee believes that examination of Alaska's estimation methodology by the MRIP's statistical consultants would be of value, particularly concerning the Access Point Angler Intercept Survey (APAIS). Staff from the MRIP and ADF&G are in contact to harmonize recreational catch accounting between the two bodies.

Texas

The Texas Parks and Wildlife Department (TPWD) manages and administers the Texas Marine Sport Harvest Monitoring Program, which includes an access-point angler intercept survey and roving counts of boats and trailers to produce estimates of private and charter boat fishing landings of finfishes. The year runs in two 6-month seasons from May 15 to May 14, and estimates are produced for each of the two 6-month periods. Annual estimates are available 6 months after year end.

The TPWD survey began 5 years before the MRFSS and was never integrated into the MRFSS/MRIP survey framework. Survey data and estimates are provided to the Gulf RecFIN database and are available to data users upon request, but they are not loaded into the regional website for data queries of marine recreational fishing catch and effort.

Unfortunately, no comparison of results between the Texas survey and the MRIP exist. Texas chose not to become part of MRFSS/MRIP because its survey was already in place when the MRFSS started, it prefers roving counts over telephone surveys for effort estimation, and its survey allows bay-specific estimates that can be used for assessment and management of state-managed species.

A full review of the Texas Marine Sport Harvest Monitoring Program is beyond the scope of this report. However, based on a presentation to the committee about the survey and on discussions with regional partners and stakeholders it is questionable whether the estimates produced by Texas are comparable to those of the MRIP. At the very least, it is highly advisable that the Texas survey be reviewed by an independent panel so that its applicability to regional fisheries assessment and management can be objectively assessed.

HUMAN DIMENSIONS

The 2006 NRC review of recreational survey methods (NRC, 2006) paid considerable attention to examining human dimension aspects of marine recreational surveys and data. That report recommended, among other things, that an independent national trip survey to support social and economic research and the continuation of add-on surveys should be developed, but designed in a way that recognizes the differences between socioeconomic and biological data to better meet management and data needs. There is a clear need for human dimension data such as demographics, angler attitudes and perceptions, expenditures, and motivations. Discussion of an independent survey to support human dimension research falls beyond the scope of this report. Human dimension add-on surveys could introduce further data-collection challenges related to such things as increased respondent burden, increased nonresponse, and item nonresponse rates that might impact stock assessment accuracy and validity. Therefore, such add-on surveys require careful consideration and, if included, should be designed to minimize data-collection problems, while meeting the need for human dimension data. Expanding the survey should be considered at length keeping in mind that such an expansion could threaten the potential reliability and validity of both types of data (e.g., effort and human dimensions). Given these challenges, further research would be needed to explore aspects of add-on survey design that will adequately address these potential issues.

CONTINUED NEED FOR A "NATIONAL PERSPECTIVE" (CONSISTENT AND COMPARABLE DATA)

Despite a deliberate focus on implementing an approach that incorporates the flexibility required to address regional and state needs, the MRIP seems to have been conscious of the overarching need for a "national perspective" for the nation's recreational fisheries surveys. Such a national view is explicitly called for in the 2006 reauthorization of the Magnuson-Stevens Fishery Conservation and Management Act (Sec. 301(a)). National standards ensure consistently high data quality that can be uniformly trusted, thereby ensuring the provision for equity among states. However, regional implementation ensures that the specific data needs of different parts of the country, each with its unique fisheries, management concerns, and priorities, can be effectively and efficiently met. As the MRIP team certifies new methods, it works with regional and state partners and stakeholders to determine how best to incorporate these methods into practice at the local level. Specifically, priority for the MRIP's support for the development and implementation of regional or state surveys is based upon the extent to which surveys, alone or in combination, can address the following criteria:

- Use MRIP-certified survey designs or methodologies;
- Achieve MRIP standards for survey coverage and basic data elements, as well as any future standards adopted by the program;
- Provide recreational catch estimates for fisheries managed under the Magnuson-Stevens Fisheries Conservation and Management Act—including Atlantic Highly Migratory Species—or jointly by the states and NMFS that are sufficient to
  - Contribute to reliable stock assessments;
  - Support development of Acceptable Biological Catch recommendations, ACLs, and accountability measures that meet Magnuson Stevens Fisheries Conservation and Management Act requirements;
  - Support development of recreational regulations that minimize triggering of accountability measures; and
  - Allow reasonably precise tracking of recreational catch against ACLs.

Therefore, this national perspective is reflected in the application of main guiding principles to survey implementation and estimation procedures. In other words, surveys that address regional and state needs might have variances or adjustments warranted by differences in regional stocks or particular fisheries, but they are structured with the same component parts (i.e., APAIS + FES) and must be based on statistically valid sampling protocols and robust estimation procedures. Adoption of these national standards ensures that, regardless of the specific decisions made by each region with respect to data-collection priorities and implementation, all recreational fisheries survey and estimation methods will withstand a rigorous independent peer review, and the resultant fisheries statistics will meet a baseline (best available science) for quality sufficient for stock assessment and fisheries management. Furthermore, implementation of the "MRIP certification" process (see Chapter 5) for acceptance of regional- or state-specific surveys provides a framework for evaluating whether these regional and state efforts meet the needed standards.

CONCLUSIONS AND RECOMMENDATIONS

Conclusion: The management landscape has changed significantly since the 2006 National Research Council report with the reauthorization of the Magnuson-Stevens Act, which mandated catch limits for all managed species. The implementation of Annual Catch Limits together with accountability measures that are enforced if the catch limits are exceeded has created additional tension in many fisheries, but particularly in recreational fisheries. Analysts, managers, and stakeholders have expressed concerns over the use of data from the MRIP to estimate catch limits and to determine whether they have been exceeded.

Recommendation: Evaluate whether the design of the MRIP for the purposes of stock assessment and determination of stock management reference points is compatible with the needs of in-season management of Annual Catch Limits. If these needs are incompatible, then the evaluation should determine an alternative method for in-season management.

Conclusion: MRIP coordination with regional and state partners has improved substantially since the National Research Council's 2006 report. In particular, substantial progress has been achieved in expanding and strengthening coordination and provision of financial, logistical, and technical support (including access to consultants) to state partners through regional Interstate Marine Fisheries Commissions and their associated Fisheries Information Networks, and the Atlantic Coast Cooperative Statistics Program. The timeliness of this support could be improved but is dependent on capacity and funding.

Conclusion: Increased communication and coordination with the Regional Fishery Management Councils and their Scientific and Statistical Committees (SSCs) has high potential to provide increased opportunities to identify and address data needs for stock assessment and management at the regional level. Furthermore, closer coordination with the SSCs would provide the MRIP with additional resources for communication and coordination with the councils.

Recommendation: The MRIP should continue and expand its investments to coordinate with, and provide financial, logistical, and technical support to, regional Interstate Marine Fisheries Commissions and state partners.

Conclusion: The MRIP has adopted a regional implementation approach that incorporates the flexibility required to address unique regional data needs. Thus, the program has evolved to become a compilation of regionally based data-collection programs and is better prepared to address data needs at regional and state levels. Continued coordination, technical support, and integration of Pacific Coast state surveys (Washington, Oregon, and California) into the MRIP framework are warranted. Despite noticeable progress on regional coordination, Pacific Coast states still perceive the MRIP as only a partial solution to their long-term recreational data-collection needs. Furthermore, flat or reduced funding has made implementation of recommended survey improvements difficult and in some cases impossible for Pacific Coast states to accomplish. If this problem persists, any meaningful gains to the states' sampling programs through the MRIP certification process are potentially at risk.

Recommendation: The MRIP should continue to support effective communication and coordination with Pacific Coast states. Coordination should be focused not only on continued logistical and technical support for survey improvements and subsequent MRIP certification, but also on better articulation of the benefits of a flexible regional approach to data collection, and interstate survey coordination for broad-scale stock assessment and fisheries management.

Conclusion: The MRIP has continued to maintain a national perspective for development and implementation of recreational fisheries data collection by establishing and maintaining a certification process for acceptance of regional- or state-specific surveys. This certification process provides a framework for evaluating how the regional and state efforts meet the basic MRIP requirements and produce outputs suitable for stock assessment and management advice.

Recommendation: The MRIP should increase efforts to clearly articulate to regional and state partners, as well as anglers and other user groups, the meaning, significance, and importance of the current approach to implement its national perspective on recreational fishing surveys. The MRIP should also be clear that this national approach incorporates the appropriate amount of flexibility required to meet unique regional and state needs. The benefits of a cohesive, integrated, and statistically robust recreational fisheries survey framework to stock assessments and regional fisheries management should be made clear.

7

Communication and Outreach with Stakeholders

INTRODUCTION

In its 2006 report, the National Research Council (NRC) recommended several improvements to outreach and communication of the Marine Recreational Fisheries Statistics Survey (MRFSS), including incorporating the views of anglers and angler organizations into the survey design and data-collection procedures and inviting them to participate in survey advisory groups. As the National Marine Fisheries Service (NMFS) developed the Marine Recreational Information Program (MRIP) to replace the MRFSS, it specifically tried to address as many of the NRC's 2006 recommendations as possible. Therefore, in its request for the current study, the NMFS asked for an evaluation of "the communication of information on survey method development, survey method descriptions, and survey results to stakeholders and application of stakeholder input in the design and implementation of new sampling and estimation procedures" (committee statement of task). This chapter provides that evaluation.

WHAT SHOULD BE COMMUNICATED AND TO WHOM

The multijurisdictional nature of combined federal and state fisheries management (see Chapter 1), which in some regions of the country necessarily involves multiple states, presents challenges to data collection, data management, assessment, and ultimately catch allocation. For example, management of summer flounder, an important recreational species along the U.S. northeast coast from Virginia to Massachusetts, necessarily involves NMFS and seven state jurisdictions that cooperate within the Atlantic States Marine Fisheries Commission (ASMFC, 2015). In contrast, fisheries for Pacific rockfishes on the West

Coast involve no more than three state agencies and one federal agency (PFMC, 2016). This differential complexity in the management of recreational fisheries is inherent to our nation's geography, but it should not be ignored in trying to understand why some recreational fisheries present greater challenges than others.

It is necessary to know specific areas of authority and responsibility within the fisheries management process to understand what should be communicated and to whom it should be communicated. The complexity of fisheries management depicted in Figure 1.2 requires that the MRIP be a component, but not the entirety of any plan for communicating fisheries management issues. Therefore, any communication plan must reflect this complexity.

There has been a substantial response by NMFS to the previous NRC report (2006), which has greatly improved the theoretical underpinnings and implementation of the elements of the MRIP. Here, we assess whether similar improvements have occurred with MRIP communication capacity and planning. The committee sought to determine to what extent the MRIP communication plan clearly identifies appropriate audiences and messaging. This assessment includes the important idea that communication is at least a bilateral endeavor involving communicating to and listening to the targeted audience.

THE MRIP STRATEGIC COMMUNICATIONS PLAN

To assist its deliberations, the committee reviewed the MRIP Strategic Communications Plan (NMFS, 2016b), copies of the MRIP's scientific and public presentations, copies of its newsletter "Newscast," and several unpublished, internal documents (NMFS, 2012a, 2012b). In addition, the committee evaluated content on the MRIP website that clearly targets an external audience. Here we focus on the 2016 MRIP Strategic Communications Plan, produced by the MRIP Communication and Education Team (CET), as it provides a comprehensive overview of current and planned activities. An earlier 2008 Communications Plan (referred to in the Sharpe and Bard presentation to the committee) was not made available to the committee, which sought to respond directly to the 2006 NRC report. Under that 2008 plan, the MRIP developed tutorial videos and a MRIP Newscast newsletter, and held several regional "road shows" (NMFS, 2012a, 2012b). In 2014, the MRIP "rebooted" its communications (Sharpe and Bard, 2016) and a new 2016 plan was developed (NMFS, 2016b).

Three aspects of the new 2016 plan are striking. First, the plan is very clear that NMFS views the MRIP as a combination of state, regional, and federal efforts rather than a monolithic federal program (Chapter 6). This aspect of the plan is appropriate and reflects the reality that the MRIP has multiple partners who play key data-collection roles. However, this begs the question of how communication about the MRIP will be coordinated among the key members of the effort.

Second, the committee was also struck by the lack of a needs assessment to identify and prioritize the current communications challenges. Elements of a needs analysis are evident in two documents that reflect on lessons learned from specific public outreach efforts (NMFS, 2012a, 2012b). However, an integrated, comprehensive needs analysis appears to be lacking.

Third, the plan is largely a vision of future communication and outreach strategies. It lacks an implementation component, which is essential given the challenge of reaching multiple partners and audiences. A few additional details are provided in the annual implementation plan updates on the MRIP website. However, it is the committee's perception that a detailed implementation plan remains to be developed, although the committee notes that the 2016 Communications Plan (NMFS, 2016b) does identify metrics for success. It simply lacks details on how and which strategies will be used to attain the metrics. Because the plan lacks important details about implementation, the committee's evaluation at this stage can only use broad brush strokes.

OBJECTIVES

The overall objective of the plan is

> to position NOAA as a trusted recreational fishing data source as well as an active and engaged partner with the appropriate expertise to lead and facilitate data collection, analysis, and reporting. The communications team works to align the *understanding* of MRIP with the *reality* of MRIP (i.e., the scope of program, the data it produces, and the uses of that data). This includes more effectively communicating and fostering dialogue on the improvements that MRIP has made to NOAA Fisheries' recreational fishing data collection processes, methods, and reporting. (NMFS, 2016b, p. 3)

The committee agrees that this overriding objective for the MRIP is appropriate. However, this broad objective focuses on external partners without explicit or full recognition of the complexity of modern fisheries management (see Chapter 1; Figure 1.2), which will require coordinated communications among the multiple partners and stakeholders in the fisheries if *understanding* is to be aligned with *reality* over broad issues in fisheries management.

The plan seeks to achieve its overall objective by implementing the following strategies (NMFS, 2016b, p. 4):

- More effectively engage partners and key stakeholders in all aspects of the MRIP program, including priority-setting, resource allocation, and implementation.
- Provide tools and resources that partners can use to more effectively communicate MRIP policies, developments, and accomplishments with their respective audiences, members, and constituents.
- Build greater awareness and understanding of the interconnectedness among data collection, science, and management functions with respect to ensuring sustainable fisheries.

- Increase confidence, enhance internal, partner, and key stakeholder understanding of the MRIP process, how it works, how it is producing more accurate and precise catch statistics, who is involved, how priorities for the program are set, and opportunities for engagement and input.
- Enhance communications, dialogue, and understanding with congressional members and staff tied to recreational data collection and program funding as part of ensuring the sustainability of recreational fishing.
- Broaden angler and for-hire operator understanding of fishing participants in the current regional recreational survey programs and increase interest and participation in those programs; recognizing that the two groups have different understandings, goals, and outreach needs.
- Improve NOAA Fisheries understanding of stakeholder's perspective on the MRIP.

The plan provides little discussion of how communication will, might, or should be differentiated, although there is a clear recognition that different audiences have different skills, needs, and other attributes. Despite the objective to "engage partners and key stakeholders in all aspects of the MRIP program," the plan focuses more on communicating to partners and stakeholders than on receiving input from the various audiences. Feedback is essential to help the MRIP effectively prioritize its future investments in human resources and technology to improve time lags between data collection and management response, which have been shown to impact management effectiveness (Sylvia et al., 2016).

AUDIENCES

The MRIP communication plan seeks to concentrate its efforts on several audiences that include internal agency partners, state fisheries agencies, the regional fishery management councils, interstate commissions, the Fisheries Information Networks, Congress, the recreational fishing community, and environmental nongovernmental organizations (NMFS, 2016b). Within this spectrum of audiences, the most effort will be focused on internal and external partners. For other key stakeholders, the CET will focus on communicating to "those that have strong, positive influence and stature among their peers" (NMFS, 2016b, p. 4).

In considering the communication plan, the committee found it easier to think of the specific roles that stakeholders play, rather than the organizations they represent *per se*. For example, a state fisheries management agency is likely to be both a data provider (agency staff may collect data from intercept interviews as a component of the MRIP) and a data user (agency staff may be involved in stock assessments and Annual Catch Target [ACT] determinations). A communication strategy that does differentiate these separate roles will likely be less effective than one that targets communication to stakeholders based on the role they represent.

Communication with Data-Collection Partners

In terms of regional operations and on-the-ground execution, the suite of surveys conducted under the MRIP umbrella is implemented in close collaboration with state and regional partners, usually by efforts coordinated through the regional Fisheries Information Networks (FINs), each of which serve a regional Interstate Marine Fisheries Commissions.[1] The regional FINs (including Atlantic Coast Cooperative Statistics Program [ACCSP]) are state-federal cooperative programs in which NMFS participates as a partner with the state fisheries agencies, interstate marine fisheries commissions, regional fishery management councils, and other federal agencies, such as the U.S. Fish and Wildlife Service. These partnerships cooperate in programs to collect, aggregate, and manage state and federal fisheries data to support fishery managers and associated agencies. The regional FINs also serve as liaisons for identifying state and regional data needs and are, therefore, well positioned to serve as coordinating entities for regional MRIP implementation. In this capacity, they effectively function as the main conduit of information and communication between the MRIP and states participating in the survey.

As noted in Chapter 6, the MRIP has made significant progress in expanding and strengthening communication and coordination with regional and state partners, especially from a logistical and survey implementation point of view (Beal, 2016; Crabtree, 2016). For example, the implementation of pilot studies to test concepts and address specific regional needs, training of state agency staff who conduct the survey, and support for database management and data systems maintenance have been successful in large part due to increased and improved communication between the MRIP and regional partners. The MRIP's broad communications strategy defers much of the communications with individual anglers or angling groups to the states and regional authorities. For example, regional RecFIN programs and state fish and wildlife agencies conduct most of the outreach and education efforts related to the Access Point Angler Intercept Survey (APAIS), effort surveys, and catch estimates, but apparently without structured and deliberate guidance from the MRIP.

Communication with Data Users

A variety of agencies, institutions, and other user groups use recreational fisheries data collected through MRIP surveys. For the purposes of this review the committee recognizes three broad categories of data customers: stock assessment analysts who use MRIP data to establish management reference points (overfish-

[1]Between 1942 and 1949, the federal government authorized by statute three interstate compacts, each creating a regional marine fisheries commission to better utilize and protect fisheries within the consenting states' jurisdiction. The three separate commissions represent the Atlantic, Gulf, and Pacific states, respectively.

ing limit and the overfished level primarily), Council Scientific and Statistical Committees (SSCs) and Advisor Panels who use MRIP data to help develop catch limits (Acceptable Biological Catch [ABC], Annual Catch Limits [ACLs], and ACT), and Council and NMFS Regional Office staff who use MRIP data to implement catch limits.

Stock Assessment Partners

The primary, direct users of MRIP information are assessment analysts who integrate estimates of catch and sometimes catch per unit effort of recreational anglers into stock assessments (see Figure 1.2). Presentations to the committee and discussions between the committee and stock assessment analysts revealed that the assessment analysts broadly recognize the improvements made to the MRIP since 2006 (e.g., Dick, 2016). Assessment analysts have found MRIP staff to be responsive to requests for the data used to develop catch limits and to be knowledgeable about the underlying data (e.g., Carmichael and Duval, 2016). In these cases, communication is typically one-on-one, often relying on personal relationships developed over years between MRIP staff and assessment analysts.

Overall, the committee encourages the MRIP to continue two-way communication with customers who use data to generate ABCs, ACLs, and ACTs. The benefits accruing to the MRIP from such communication would include improved understanding of how data are used in the assessment process, enhanced effectiveness of survey query tools, and prioritization of design improvements. In addition, the assessment analysts would gain a better understanding of the strengths and weaknesses of the MRIP data. Finally, although this communication will largely continue to be peer-to-peer, the MRIP should seek opportunities to engage assessment analysts in group situations wherever possible.

Partners Who Establish Catch Limits

Since the 2006 report, the Magnuson-Stevens Act reauthorization assigned specific responsibility to the SSCs of each regional management council to set ABCs for management species.[2] Because they are required to allow for scientific uncertainty in developing their catch recommendations, the SSCs have become an important new user of MRIP information. More so than assessments analysts, who may need to limit the uncertainty in catch estimates to an arbitrary low figure for analytic purposes (e.g., 5 percent for red snapper in SEDAR, 2013; Boreman, 2016), the SSCs must fully consider the uncertainty in catch and catch per unit effort (CPUE) estimates. According to presentations to the committee (Boreman, 2016; Dick, 2016), engagement of the SSCs by the MRIP is in the early stages

[2] Magnuson-Stevens Fishery Conservation and Management Reauthorization Act of 2006, 16 U.S.C. § 1852(g).

and requires increased emphasis. Groups with responsibilities similar to those of the SSCs also exist within the fishery commissions and sometimes within individual states (Beal, 2016). These groups would also benefit from improved communication with the MRIP.

The bodies that subsequently establish ACLs and ACTs must account for management uncertainty rather than scientific uncertainty. Here, too, improvements in how the MRIP communicates with these bodies would be helpful in aligning the understanding of recreational fisheries with the reality of managing them.

Finally, there is a separate category of data users that regulate transboundary resources, for example, the Pacific Salmon Commission and the International Pacific Halibut Commission (IPHC). For these bodies, historical transmission of recreational catch data for stock assessment purposes has been through state agencies. Information transfer from the MRIP to these bodies has not occurred for two reasons. First, the major source of information for recreational catch of these species has been the Alaska Department of Fish and Game, which has not traditionally participated in the MRIP. Second, the IPHC does not itself conduct any in-season management of recreational halibut catch limits; instead, it relies on agencies of the contracting parties for such management.

Partners Who Set and Enforce Catch Limits

Based on the evidence presented to the committee, the biggest single challenge confronting the MRIP is the use of its data in the management arena—specifically in implementing catch limits (Figure 1.2). This is a new requirement of the Magnuson-Stevens Fishery Conservation and Management Reauthorization Act of 2006. The MRIP collects data at specific sites and times through complex surveys that collectively produce estimates of fishing effort and CPUE. As outlined in Chapter 2, these data collected at specific places and times are aggregated into broader spatial areas and into 2-month waves (Box 2.2). Moreover, for most stock assessment purposes, these data are further aggregated to annual estimates for a single region. For this application of the MRIP, the effects of lags in data entry, data processing, quality assurance and quality control procedures, and reporting inherent in this aggregation are relatively minor, because data used for assessments are typically from the previous complete fishing year (Sylvia et al., 2016). However, in sharp contrast, when MRIP data are used to enforce catch limits, inferences are made for limited spatial areas (e.g., states) and at very specific times (e.g., on what day the ACT was or will be reached). This creates a tension in the fisheries management process in which a survey designed for one purpose is being used for another purpose, for which its design may not be optimal. Moreover, because of the need for regional management councils to avoid overfishing, the uncertainties associated with catch estimates become of critical importance—possibly leading to fisheries closures long before the point estimate is reached to ensure that ACLs are not exceeded (Baum, 2016). Regional managers also expressed frustration over the timeliness of MRIP estimates for implementing catch limits—particularly when fisheries are pulsed as is the case for many recreational fisheries (Carmichael and Duval, 2016; Crabtree, 2016). There was also concern expressed, and perhaps a lack of understanding, over the estimates of uncertainty (proportional standard errors) that accompany the individual catch estimates (Carmichael and Duval, 2016; Crabtree, 2016).

In general, the committee found that communication challenges continue to exist regarding data users (Baum, 2016; Boreman, 2016; Crabtree, 2016). For example, perceptions of poor communication in the Gulf have generated support for development of a Regional MRIP Communications Committee within the Gulf States Marine Fisheries Council (GSMFC) RecFIN structure, to be populated by a GSMFC staff member plus representatives from each of the five Gulf states.

COMMUNICATION WITH AFFECTED USERS

Effective communication with affected users is clearly essential to ensure maximal cooperation with MRIP survey instruments, compliance with the fishing regulations that are subsequently derived from MRIP estimates, and support for the MRIP program overall. Through oral presentations (Carmichael and Duval, 2016; Crabtree, 2016; Martin et al., 2016; Mumford, 2016) and written submissions, the committee heard that there remains scope for substantial improvements in communications about the MRIP generally, and its role in the overall setting of catch limits and in enforcing catch limits specifically.

Recreational Anglers

The MRIP has indicated that it expects most of the effort to communicate with anglers will be conducted through its state and regional partners (Sharpe and Bard, 2016). Whether it works through partners or directly, a major issue for anglers is the credibility of the MRIP data-gathering process and the data themselves. The most common issues heard by the committee were that many anglers have rarely, if ever, been surveyed, that anglers wonder why the APAIS is conducted at times and places where the fishing is poor or there are few anglers, and that the data reported by the MRIP do not conform to their individual experiences (e.g., Sharpe and Bard, 2016). Anglers also shared that they are catching a lot of fish but that the MRIP is reporting low catch rates, or vice versa. Many possible reasons for these impressions exist—some of which can be explained from a first-principles understanding of the survey. For example, a statistically valid sample of millions of anglers will likely still miss most of the anglers who are actually fishing, giving them the false impression that the survey is inadequate. In addition, an angler's individual experiences might not match the collective experience of all anglers, leading to the perception that the MRIP data missed that individual angler's experience. Although these and other fundamental concerns over the design and implementation of the MRIP may be easily explained, they are real perceptions by recreational anglers that can only be addressed by proactive engagement of these key stakeholders on multiple fronts. The committee heard that communication shortcomings have exacerbated these concerns to the point where some stakeholders question the value of the MRIP in ensuring sustainable management of the nation's marine living resources (Carmichael and Duval, 2016; Martin et al., 2016). Addressing these concerns is a primary concern. The committee recognizes that an approach coordinated with the states may be most successful in building trust and aligning the understanding of these stakeholders with the reality of how the MRIP is deployed. However, the MRIP must play a leading role in providing the vision and implementation strategies that partners can follow to ensure that affected users regain confidence in the MRIP's data products.

For-Hire Sector

The for-hire sector, which includes charter boats, headboats, and guided small boats (e.g., flats and skiffs), is in effect a commercial sector, but unlike other commercial fisheries it makes its profits from charging clients rather than selling fish themselves. Many in the for-hire sector hold strong views regarding fisheries management in general, and on the MRIP and its predecessor MRFSS in particular. For-hire captains tend to be better informed about fisheries and fisheries management issues than individual anglers, and often (but not always) are willing to cooperate in data collection. Many are directly involved in the fisheries management process because fishing is not recreation for them, but their livelihoods. They tend to belong to associations, and they tend to have a broader view of their sectors of recreational fisheries than do individual anglers.

The potential for for-hire captains to be fully cooperating partners is likely greater than that for individual anglers. These partners are more likely to be motivated to actively seek out communications from the MRIP. Although the issues of concern to them are similar to those of individual recreational anglers, they often have different experiences and knowledge bases. As a result, the style and content of MRIP communication to this sector should differ from that targeting individual anglers. Currently, the MRIP has communication products aimed directly at this group (e.g., MRIP Angler Catch Surveys: Information for Marinas and For-Hire Operators[3]). These directed communication efforts should be continued because of both the direct benefits from an engaged for-hire sector and the indirect benefits from the operators' interactions with clients. Furthermore, the MRIP should provide the for-hire captains with a method to review *their own* data submittals to provide further quality assurance of these data. The committee recognizes that the

---

[3] See http://www.st.nmfs.noaa.gov/Assets/recreational/pdf/FINAL-2016-Marina-Charter-Boat-Handout.pdf.

MRIP must follow federal regulations to maintain data privacy and anonymity. The committee also recognizes that this additional step for data submittal would assuage concerns of an important fishing sector about the quality and accuracy of their own data.

The for-hire sector expressed concerns to the committee about delays in adopting self-reporting of catch through mobile or web-enabled devices (e.g., Brennan, 2016). As noted in Chapter 4, there can be serious statistical design issues for self-reporting, whether by individual anglers or charter captains, related to avidity, reporting bias, and representativeness of reporting that must be addressed. However, the obvious advantages, including timeliness, auditing, and gaining of longitudinal data on variation and characteristics of catch, could improve the management process. Therefore, the committee encourages the MRIP to address the statistical issues associated with self-reported data. Some pilot programs are being conducted to evaluate the feasibility of such options, but the pressure from stakeholders to adopt self-reporting systems can be expected to increase. From a communications viewpoint, this pressure can be mitigated by ensuring that materials that explain the statistical issues associated with self-reported data are a priority in the MRIP communication plan. Though the for-hire sector has expressed a particular interest in the implantation of electronic self-reporting options, individual anglers may also benefit from this material.

Other Recreational Fishing Businesses

Relevant businesses include marinas, bait and tackle and fly-fishing shops, boat-rental operations, fishing piers, general sporting-goods stores, local grocery stores and restaurants, and others. The degree of interest in the MRIP and the potential for involvement as partners both vary depending on the degree to which businesses are directly involved in recreational fishing. Thus, marinas and tackle stores are likely to be more involved than grocery stores and restaurants. However, leveraging of the potential of these commercial operators will require directly tailored communication products.

COMMUNICATION STRATEGIES

The MRIP communication strategies continue to evolve. As this occurs, the committee suggests that it will be important to distinguish between passive and interactive communication activities. Passive and one-way activities involve the passing of information from one entity to another and take the form of such media as newsletters, magazines, posters, pamphlets, and websites, to name a few. Because this mode of communication relies on people choosing to engage with the MRIP, it necessitates that communication materials be visually appealing, clearly written, and devoid of jargon. This is particularly important in the case of the MRIP, given the need to convey rather complex scientific and statistical information to a generally nonscientific constituency. The materials must be written and presented in a way that piques interest and increases trust in the overall enterprise. Nonetheless, although good, clear passive communications are an essential part of any communications plan, they are not sufficient.

The concepts of active communication and outreach are used almost synonymously throughout the MRIP Communications Plan. The committee does not attempt to distinguish between the concepts here, but instead discusses aspects of communication and outreach that appear to be critically important, whatever they are called. Interactive communications have the intent of reaching a broader audience using communication methods that are two-way and involve more give and take. Activities of this form include presentations at public events, workshops, and booths at recreational fishing shows, to name a few. In addition, interactive communications are often realized by including interactive features on a website. Thus, for example, the website could invite users to share their thoughts, offer opportunities to partner in data collection, offer opportunities to provide input about the communications plan and other aspects of the MRIP, and so on. The MRIP Communications Plan contains elements of both one-way and two-way communication, but there is little in the way of specifics, particularly with respect to two-way communication. As stated earlier in this chapter, the plan focuses primarily on communication from the MRIP to partners and stakeholders with little to no discussion of efforts regarding collaboration, feedback, and input from the various audiences and methods for engaging with partners and stakeholders more interactively.

Four principal modes of communication can be recognized within the MRIP Communications Plan (NMFS, 2016b).

The MRIP Website

The MRIP website (http://www.st.nmfs.noaa.gov/recreational-fisheries/index, also reachable at http://www.countmyfish.noaa.gov) is well laid out, reasonably easy to navigate, and extremely informative. It provides information on survey methods, fishery data collected through the MRIP, documentation, and other aspects of the program. The material is presented under major categories, each with its own drop-down menu. The committee judges the information to be detailed and likely accessible to any interested person. NMFS appears to be developing audience-specific navigation pathways on the website to help users find information at the appropriate level of technical detail. As noted above, the committee also encourages the MRIP to use web-based technologies to stimulate active, two-way communication on its website.

The committee recognizes that the surveys and communication materials have gone through many rounds of changes and improvements. However, the committee has struggled to locate detailed information on the technical basis for current estimations and procedures. This information may not be sought by all audiences, but groups interested in the technical details of the survey (e.g., state partners, the FINs, ACCSP, as well as some individual anglers and angler groups) would likely benefit from ready access to this information. It also would be helpful to include the date it was produced and whether there is a more recent version currently in use. Thorough documentation of the statistical practices would support continued evaluation and improvements to the system.

The MRIP Newscast

The MRIP Newscast newsletter has been produced since 2008 and is available electronically for interested stakeholders. The Newscast provides a range of topical information related to the MRIP and recreational fisheries generally. The standard and information content of the newsletter is high. Because the newsletter is delivered electronically, the CET plans to use the web (e.g., number of opens, click-through rates, points of origin) to understand and presumably target its audience better (Sharp and Bard, 2016).

Engaging Data-Collection Partners and Data Users

A component of the 2016 MRIP Communications Plan (NMFS, 2016b) calls for engagement of data-collection partners and data users at Council and Commission meetings to build understanding of the changes and improvements to the MRIP. As mainly data-collection partners and constituents who use the MRIP data to enforce ACLs and ACTs, these are appropriate audiences to engage. Details as to how this engagement will occur were largely lacking, and engagement of constituents who use the data to generate reference points and ABCs was not described.

Print and Social Media Products

The 2016 Communications Plan proposes to continue to produce both traditional and social media products to explain forthcoming changes to the MRIP, but few details were provided.

STAFFING THE MRIP CET

The 2016 Communications Plan identifies a two-tiered structure (NMFS, 2016b; Sharpe and Bard, 2016) involving a national team and regional teams. The national team will consist of the MRIP communication staff (two full-time equivalents) and representatives from other National Oceanic and Atmospheric Administration offices and regional interstate commissions. Similarly, regional teams will consist of representatives from the states, councils and commissions, and other key regional partners. This structure appears a reasonable one to develop and implement a communication plan that emphasizes regional connections and differences (NMFS, 2016b). The extent to which the regional teams have been created is unclear.

The committee recognizes the challenge with undertaking the broad portfolio of communications discussed in the 2016 Communications Plan with current staff. The MRIP has made excellent use of statistical consultants to respond to the previous NRC review of recreational fishery surveys. The MRIP has already begun to use a communications consulting firm. Further benefits may accrue if the MRIP works more closely with external communication professionals to help them develop and implement their communication plan moving forward.

CONCLUSIONS AND RECOMMENDATIONS

Conclusion: Fisheries management in state and federal waters is a complex, multidisciplinary, multistakeholder process. Communications about the MRIP must be undertaken in the context of the entire fisheries management process. The MRIP Communications Plan alone cannot be expected to explain to stakeholders all the complexities of fisheries management. The MRIP Communications Plan cannot and should not be expected to communicate to all audiences the comprehensive nature of fisheries management.

Recommendation: NMFS should develop and lead an integrated communications strategy involving state and federal partners to explain and seek support for the management of the nation's fisheries within which the role of the MRIP is clearly defined. The MRIP Communications Plan should be an element—albeit for species in which removals are dominated by recreational fisheries, an essential component—of such a broader, integrated overall communications plan.

Conclusion: The MRIP Communications Plan lacks a clear needs analysis and an implementation plan. The plan identifies broadly what the MRIP wishes to achieve, but there is little discussion of specific and practical matters that the MRIP and its predecessor, the MRFSS, from which the MRIP must have learned. The plan lacks details about specific and practical matters such as where to place information, what outlets to use for different kinds of information (e.g., newspapers, angling magazines, local television, tackle shops), and how to ascertain what MRIP users and those affected by the MRIP think of the plan and what they would like to learn more about. It is not enough to produce a detailed, extensive, and informative website, no matter how good that website might be.

Recommendation: The MRIP should further develop its communications plan to include a specific needs analysis and develop a specific and detailed implementation plan. Greater emphasis should be placed on interactive (two-way) communication, which may involve spending time in the field with anglers, than is currently in the plan.

Conclusion: The MRIP Communications Plan identifies a hierarchical structure with both national and regional teams. This seems appropriate given the regional-state and federal nature of the MRIP partnership. In response to concerns regarding statistical aspects of the survey expressed by the National Research Council (NRC, 2006), the MRIP established a large team of statistical experts, both in-house and as consultants, to help with the redesign of its sampling methods, analyses, and surveys. A similarly experienced team of experts has not been established to support the MRIP communication and outreach activities.

Recommendation: The MRIP's success depends to a large degree on clear, accurate, and timely communications, and on engaging all the various stakeholder groups, including anglers. Therefore, whether as permanent full-time equivalents or as consultants, the MRIP should consider expanding its communications team to support the required needs analysis and implementation plans identified by the committee. One way of achieving this expansion would be to partner with national and regional organizations, such as the Sea Grant colleges, that already have communications capacity and expertise and could identify opinion leaders and constituencies.

Conclusion: The MRIP has made significant improvements to its communications and outreach strategy since the National Research Council's 2006 report. Perhaps the most significant improvements have been to its website and communications with some of its data-collection partners, such as the regional interstate marine fishery commissions and state fishery agencies. Its communications with some other groups, most notably anglers, but also some stock-assessment and management groups, have been less successful. Significant communications challenges remain unaddressed.

Conclusion: There is a need for increased and regular coordination and communication with regional fishery management councils and their scientific and statistical committees, and with the regional stock assessment programs. This increased communication would provide opportunities for identifying and addressing data needs for stock assessment and management at the regional level.

Recommendation: NMFS should develop a system for indexing and cross-referencing documentation of survey methods and statistical analysis. Because of the evolving nature of the program that includes many different elements, maintaining the organization of the technical documents is a challenge. NMFS should increase its efforts to ensure that documentation includes key pieces of information. For example, NMFS should ensure that the statistical basis for the stratified and total estimates of total effort, catch per unit effort, and their variances for all fisheries and areas are readily available and consistent among current documents.

Recommendation: The MRIP should take a more active role in communicating with anglers, whether through its partners or through its own efforts. The committee recognizes that the MRIP defers to the states and regions in communications with anglers. Furthermore, the committee recognizes that an approach coordinated with the states may be most successful in building trust and aligning the understanding of these stakeholders with the reality of how the MRIP is deployed. However, the MRIP should play a leading role in providing the vision and implementation strategies that partners can follow.

Recommendation: The MRIP should provide the for-hire captains with a method to review *their own* data submittals to provide further quality assurance of these data. The committee recognizes that the MRIP must follow federal regulations to maintain data privacy and anonymity. The committee also recognizes that this additional step for data submittal would assuage concerns of an important fishing sector about the quality and accuracy of their own data.

8

Plans for Maintaining Continuity

INTRODUCTION

The Marine Recreation Information Program (MRIP) developed improved methodologies for the estimation of both fishing effort (the Fishing Effort Survey [FES]) and catch per unit effort (the Access Point Angler Intercept Survey [APAIS]) by recreational anglers. The resulting estimates of catches differed from those produced by Marine Recreational Fisheries Statistics Survey (MRFSS) and created the need to link and calibrate previous information collected under MRFSS with the new information from the MRIP, to create a continuous time series of equivalent data. In this chapter, the committee describes and evaluates the methods developed for this linking process and its implications for the assessment and management of stocks utilized by recreational anglers.

THE NEED FOR CONTINUOUS DATA SERIES

The MRIP calibration workshops, presentations to the committee, and substantial public testimony have highlighted the need for continuity in the recreational fisheries data used for assessment, management, and allocation. The three different processes have differing capabilities to accommodate changes in historical estimates. The stock assessment process can use recreational catch and effort statistics in two ways: as part of the raw data inputs on removals, and as indices of relative abundance. Changes in time series resulting from design and estimation changes can generally be accommodated inside assessment models using temporal blocks with different catchabilities for the two components of the time series (MRFSS/MRIP). The alternative approach is to calibrate the two time series to each other external to the assessment model and use the calibrated estimates directly. Each approach has its merits, and while the internal assessment model treatment is more robust to uncertainty, the need for a common time series to use in other applications (management, allocation) argues in favor of the external calibration approach.

For assessment and management programs where there is no statistical model used for assessment and where the Annual Catch Limits (ACLs) may be based on historical trends, the calibrated approach is essential as a consistent yardstick for calculation of long-term averages and their variance. The MRIP calibration workshops clearly identified that modifications of the survey methodology required historical estimates to be calibrated to current methodology, rather than the opposite. The implications of an adjusted time series of catch estimates could be significant in the allocation arena, and some aspects of this issue are detailed in following sections. Likewise, adjusted time series of catch or effort statistics can influence the development of control rules for fishery removals. For example, calculation of season lengths or bag limits designed to maintain historical angler success or access will be sensitive to the input data. Existing control rules used for input management control may need to be reassessed in light of the adjusted time series of catch estimates by time or area.

TRANSITION FROM PHONE-BASED TO MAIL-BASED EFFORT SURVEY

The Coastal Household Telephone Survey (CHTS) was an extremely problematic element of the MRFSS due to a number of potential and realized biases in a methodology based on random-digit dialing of landlines. The 2006 National Research Council (NRC) report noted the inherent difficulty of estimating fishing effort using such methodology, in the absence of an adequate list frame of anglers to increase the efficiency and accuracy of the effort estimation. The MRIP has clearly heeded the NRC advice and developed a dual-frame methodology using both a list frame of anglers and a secondary list frame based on the U.S. Postal Service address-based frame of households (Chapter 3). The MRIP undertook substantial design and testing of the new effort estimation methodology. The results of implementing the new procedures were different estimates of fishing effort, often by large amounts, for some areas and time periods. Andrews et al. (2014, Table 3, p. 18) document differences in fishing effort of approximately four times higher for the improved FES compared with the previous CHTS methodology. Because these estimates resulted in much higher estimates of total catch for species in these areas and times, the committee has invested considerable effort in examining their validity.

Chapter 3 of this report examines the MRIP effort estimation methodology in detail and makes several recommendations to address issues of nonresponse and recall biases, weighting of the strata responses, and correct incorporation of variance in the components of the ultimate estimates. These recommendations are significant to the estimation of fishing effort, catch per angler, and the ultimate calculation of accurate values for total catch. Although addressing these recommendations may change the scale of the MRIP time series of total catch amounts for some areas and species, the choice of a method for calibration of the MRFSS and MRIP time series is not likely to be sensitive to these changes. This is because the changes contemplated by this report will affect primarily the degree of offset between the two time series, which the calibration is designed to bridge. However, it is important that MRIP staff be cognizant of any changes in methodology that affect the determination of peak fishing effort periods, because all calibration methods currently contemplated involve the use of peak effort periods to calibrate MRFSS estimates to MRIP estimates.

DEVELOPMENT OF CALIBRATION AND BRIDGING AMONG DATA SERIES

The 2006 NRC report on the MRFSS program (NRC, 2006) recommended several improvements to the program that would reduce the potential for bias in both effort and catch estimation. Largely as a result of the NRC report, the National Marine Fisheries Service initiated a complete redesign of both the effort and catch components of the MRFSS program. The two elements of the MRIP (APAIS and FES) were implemented with different degrees of rigor, largely dictated by the relatively higher expense of the intercept survey. The APAIS was evaluated in a side-by-side comparison with the previous MRFSS methodology in only a single year and for a single area. As such, our knowledge of the relationship of the estimates arising from the two methods is somewhat limited. In contrast, a carefully staged implementation of an improved mail-based FES was more temporally and spatially extensive. Nonetheless, the combined MRIP methodologies resulted in estimates of recreational catches that differ from the previous MRFSS estimates, generally by small amounts but substantially for some species-area units (Andrews et al., 2014). These differences between MRFSS and MRIP estimates ranged from consistent biases to apparently random variation.

The new methodologies employed in calculating the MRIP estimates are more statistically valid than those used in the MRFSS program (Chapters 3 and 4). Many important components of recreational fisheries management are dependent on these catch and effort estimates, including stock assessment, development of harvest policies, in-season management, and catch allocation (Figure 1.2). In addition, the allocation of resources for the production of catch statistics is itself dependent on the estimates of catch produced by the MRIP. The historical time series of recreational catch and effort produced with the outdated MRFSS procedures therefore requires calibration to the estimation processes used in the MRIP, so that a combined time series of total removals may be used to inform these processes.

The MRIP convened two workshops to address the calibration issues. The first, in 2012, was designed to develop a method to calibrate 2004-2011 catch rate estimates based on the unweighted MRFSS estimation methods to catch estimates based on a new MRIP weighted method, demonstrate its use in hind-casting estimates prior to 2004, and develop a plan for implementing the calibration into benchmark stock assessments. The workshop identified a simple ratio estimator (MRFSS/MRIP) using 2004-2011 data, which could be used as a constant for hind-casting data prior to 2004, or trended using auxiliary information. The second workshop, in 2014, was convened to revisit the calibration issue in light of changes to the APAIS made in 2013 and 2014. That workshop identified three potential alternatives for calibration (discussed below), an interim methodology to use while the three methods were evaluated fully, and procedures to follow if survey methodology were to change in the future.

Both workshops clearly recognized that calibration was critical in allowing stock assessments to differentiate true changes in stock status from changes in the estimation procedures producing the data used in the assessments. Both workshops also identified several issues that affect the sampling error of the catch estimates, based on changes to the survey designs of both the MRFSS and MRIP over time.

The committee reviewed the workshop reports and other MRIP documents to determine the current status of calibration and plans for updating or improving the calibration method. Appendices 1 and 2 of the 2014 Calibration Workshop report (Carmichael and Van Voorhees, 2014) outline the three suggested alternatives for calibrating pre-2013 estimates to the post-2013 estimates. Importantly, the workshop also considered the opposite calibration, that is, calibrating the post-2013 estimates to the historical time series. The workshop concluded that the former process (calibrating historical to present) was the preferred calibration method because harvest control methodology requires coherence with catch estimation methodology.

The three alternative methods were examined thoroughly by the 2014 workshop. Their construction and merits are detailed in Appendix 1 of the workshop report, and are only summarized here (Carmichael and Van Voorhees, 2014).

1. Direct catch ratio estimator. In basic concept, the simple ratio estimator takes advantage of the improved coverage of peak periods in the 2013 MRIP ($C_{p,\,2013}$) and scales the catches prior to 2013 by the ratio of peak catches to total catches in 2013 ($R_{2013} = C_{total\ 2013}/C_{p\ 2013}$). The scaled estimate for total catch ($C_{tot,y}$) in prior year *y* is then based on applying the 2013 ratio to the peak catch in the prior year, *y*. Thus,

$$C_{tot,\,y} = R_{2013}{*}C_{p,y}.$$

   The scaling is based on post facto identification of peak periods prior to 2013 and makes no use of data for nonpeak periods.
2. Complex ratio estimator. Because the MRIP program produces estimates of effort distribution throughout the day, it provides an opportunity to

scale the effort distribution in 2013 to match the truncated effort estimate from the more limited sampling in prior years. This is achieved by adjusting the weighting of effort in temporal bins for 2013 to match the effort for the more restricted temporal bins that existed in previous years. The ratio of the catch in the truncated 2013 bins ($C_{tr\ 2013}$) to the total catch in 2013 (C_{2013}) is defined as $R_{c/tr,\ 2013}$. Similar to the simple ratio method, the $R_{c/tr,\ 2013}$ is then applied to the available catch estimate (C_y) from a prior year *y*, to obtain an adjusted estimate of catch for that year. Thus,

$$C_{y,\ adj} = R_{c/tr,\ 2013} * C_y.$$

This adjusted estimate is assumed to be the estimate that would have been obtained if more complete MRIP-style sampling had been conducted in previous years. This estimator assumes a constant distribution of catch and effort over time and area, relative to that in 2013. Furthermore, estimates of total effort for years prior to 2013 are obtained from the CHTS, which have unknown properties.

3. The regression-based estimator. This estimator is more involved than the ratio estimators and is in some measure the reverse of the complex ratio estimator. It uses 2013 data to estimate and predict the distribution of morning, peak, and evening categories of catch/effort for 2013, based on characteristics of the catch or demographics from the APAIS. This modeled relationship is then applied to target year data to derive a pseudo distribution of categories for that year, which matches the 2013 distribution. These pseudo proportions are used to produce adjusted estimates of catches for the target year. Several extensions to this method are outlined in the report.

   The primary assumption of this method, and it is a strong one, rests on the stationarity of the catch and effort process over time and space. In other words, it assumes that the effort and catch distribution throughout any given day can be captured by this single model relationship. The committee appreciates the conceptual investment in this approach and commends the workshop for its innovative thinking. However, the committee has strong concerns about the ability to validate such an approach, because the quantity being predicted, that is, the distribution of categories, cannot be observed. This quantity is defined by 2013 characteristics and imputed to the target years.

The committee notes that all three methods are actually model-based estimators—all involve an underlying estimation model and vary only in the influence of the assumptions involved for each. The workshop consultants recognize that the calibration was not straightforward due to the limited side-by-side estimation using previous and current methodology for almost all areas. The committee agrees with the consultants' concern in this regard and believes that uncertainty about process and observation error could be reduced if additional side-by-side comparisons were conducted. While the consultants also suggested that time-series or small-area spatial analyses might also be conducted, the committee is doubtful that such analyses would yield significant improvements in a general calibration method. Nonetheless, such analyses could be conducted with available data and would be worth some investment of analytical resources.

Appendix 2 of the 2014 workshop also identified an interim approach (the simple ratio) to be applied while a full evaluation of the three alternatives was conducted. The Appendix detailed the drawbacks to this method, notably that the relationship of peak period catch to total catch is constant, and that none of the data outside of the peak catch period for years prior to 2013 are used. Both the 2012 and the 2014 calibration workshops provided guidance to stock assessment scientists concerning the use of a calibrated time series for the combined MRFSS-MRIP data. In particular, they suggested increasing the assumed variance in the time series to account for uncertainty in the calibration process.

ANTICIPATING IMPACTS ON ASSESSMENT AND MANAGEMENT PROGRAMS

An accurate calibration of MRFSS data to MRIP data has implications for both assessment and management. Statistical catch-at-age (SCA) stock assessments, while not immune to differences, are the least sensitive to calibration issues because the assessment models can accommodate some imprecision in calibration through alternative catchability functions. Imprecise or biased calibration does affect the calculation of reference points related to unfished biomass, hence optimum harvest rates and control rules. In SCA frameworks, calibration issues may increase uncertainty in these quantities, although these influences will be less strong than in other assessment/management frameworks.

In non-SCA stock assessment frameworks, and particularly in data-poor assessments, where the time series of total catch is a prime determinant of harvest levels (ACLs) and reference points, the method of calibrating MRFSS to MRIP data is likely to be more influential than in SCA frameworks. In the former, both the trend and scale of stock changes are informed totally by the calibrated time series, and in turn, the understanding of stock status is similarly governed. In these instances, the calibration process will have a much larger influence on the understanding of current stock status and appropriate reference points for stock management. The committee notes that these influences will not be uniform and will affect recreational fisheries management much more strongly in some areas than others, directly linked to the nature of how ACLs are determined.

For data-poor assessments the estimation of common reference points for stock management, for example, unfished equilibrium biomass B_0, biomass depletion level, and target harvest rate, are not well determined, or may be precluded, by time series of catches alone. The estimated B_0 is a quantity of considerable uncertainty for even technically sophisticated assessments. In turn, a poor understanding of target harvest rates will increase the uncertainty associated with ACLs. Calibration affects primarily the scale of estimated removals but may also influence its trend. In the absence of auxiliary information on trend, management is therefore critically dependent on simple catch time series. These issues are not uniquely associated with the methodology for calibrating data series resulting from changed estimation methodologies, but imprecision in calibration will increase uncertainty in fisheries management.

Future efforts to develop calibrated time series of recreational catches will be most useful if accompanied with advice on the implications of the calibration method to stock assessment and reference points for stock management. In particular, simulation analyses of alternative methods will be useful. As the time since a change in methodology for estimating recreational catches lengthens, the calibration method will have less influence on the understanding of current stock status. The understanding of stock status will be influenced more strongly by recent data than by historical shifts in estimation methodology for catch, when removals are substantial proportions of available yield. If removals are a small proportion of available yield, then the calibration will continue to influence understanding of stock status. However, because the calibration methodology does influence the understanding of reference points for management, the effect of the calibration will be a long-term element of fisheries management. This is an important element to consider when contemplating any changes in survey and estimation methodology and underscores the point that any such change should be thoroughly evaluated prior to implementation.

CONCLUSIONS AND RECOMMENDATIONS

Conclusion: The low number of side-by-side comparisons between the angler intercept portions of the MRFSS and MRIP methodologies limits the ability to develop a more precise calibration between the time series of data produced by the two programs.

Conclusion: None of the methods proposed to calibrate the two intercept time series is completely satisfactory because of the necessary assumptions and/or post hoc data stratifications that must be applied when using the methods.

Conclusion: For stocks with substantial removals, the calibration between the two intercept data sets will diminish in importance for some stock assessment purposes over time as more recent data dominate the determination of stock status. Nonetheless, uncertainty in the estimation of reference points for harvest policy determinations will remain sensitive to the calibration process.

Conclusion: The calibration of the two data time series is extremely important to multiple aspects of fishery stock assessment, catch management, and allocation processes. For stock assessment modeling, the absence of a fully satisfactory calibration can be addressed through alternative estimates of catchability over the combined time series. For simpler stock assessments, the calibration may be more influential.

Recommendation: The MRIP should continue development of a statistically sound calibration methodology as improvements to the Access Point Angler Intercept Survey and Fishing Effort Survey methodologies are incorporated. In the interim, the existing ratio-based calibration should be continued. For statistical catch-at-age (SCA)-based assessments, scientists should employ alternative catchability functions applied to the combined time series as a means to accommodate potential imprecision in the calibration of MRFSS data to MRIP data. For non-SCA frameworks, assessment scientists should exercise caution in the interpretation of trends in catch data.

References

AAPOR (American Association for Public Opinion Research). 2016. Address-based sampling. Report prepared for AAPOR Council by the Task Force on Address-based Sampling (R. Harter, Chair). http://www.aapor.org/getattachment/Education-Resources/Reports/AAPOR_Report_1_7_16_CLEAN-COPY-FINAL-(2).pdf.aspx.

Anderson, M. 2015. Technology Device Ownership: 2015. Pew Research Center Report on Internet, Science and Tech, October 29, 2015. http://www.pewinternet.org/2015/10/29/technology-device-ownership-2015/.

Andrews, R., J. M. Brick, and N. A. Mathiowetz. 2014. Development and Testing of Recreational Fishing Effort Surveys: Testing a Mail Survey Design. NMFS Final Report, July 31. https://www.st.nmfs.noaa.gov/Assets/recreational/pdf/2012-FES_w_review_and_comments_FINAL.pdf.

Ashford, J. R., C. M. Jones, and L. Fegley. 2010. Private waterfront householders catch less per trip than other fishers: Results of a marine recreational survey. *Transactions of the American Fisheries Society* 139:1083-1090.

Ashford, J. R., C. M. Jones, L. Fegley, and R. Reilly. 2011. Catch data reported by telephone avoid public access bias in a marine recreational survey. *Transactions of the American Fisheries Society* 139(6):1751-1757.

Ashford, J. R., C. M. Jones, and L. Fegley. 2013. Independent estimates of catch by private and public access fishers avoid between-group sources of error in a marine recreational survey. *Transactions of the American Fisheries Society* 142(2):422-429.

ASMFC (Atlantic States Marine Fisheries Commission). 2015. 2015 Review of the Atlantic States Marine Fisheries Commission Fishery Management Plan for the 2014 Summer Flounder Fishery. Arlington, VA: Atlantic States Marine Fisheries Commission.

Baum, T. 2016. New Jersey Department of Environmental Protection. Presentation to NAS Committee. Washington, DC, February 25, 2016.

Beal, R. 2016. Marine Recreational Information Program. Presentation to NAS Committee. Washington, DC, February 25, 2016.

Beaumont, J. F. 2008. A new approach to weighting and inference in sample surveys. *Biometrika* 95(3):539-553.

Benaka, L. R., L. Sharpe, L. Anderson, K. Brennan, J. E. Budrick, C. Lunsford, E. Meredith, M. S. Mohr, and C. Villafana. 2014. Fisheries Release Mortality: Identifying, Prioritizing, and Resolving Data Gaps. NOAA Technical Memorandum NMFS-F/SPO-142. Silver Spring, MD: U.S. Department of Commerce, National Oceanic and Atmospheric Adminstration.

Benaka, L. R., L. Sharpe, K. Abrams, M. Campbell, J. Cope, F. Darby, E. J. Dick, J. Hyde, B. Linton, C. Lunsford, D. Rioux, and Y. Swimmer. 2016. Action Plan for Fish Release Mortality Science. Silver Spring, MD: U.S. Department of Commerce, National Oceanic and Atmospheric Adminstration.

Biemer, P. P. 2010. Total survey error: Design, implementation, and evaluation. *Public Opinion Quarterly* 74(5):817-848.

Biemer, P. P., and L. E. Lyberg. 2003. *Introduction to Survey Quality*. Hoboken, NJ: John Wiley and Sons, Inc.

Blumberg, S. J., and J. V. Luke. 2015. Wireless Substitution: Early Release of Estimates from the National Health Interview Survey, January–June 2015. National Center for Health Statistics, U.S. Department of Health and Human Services report, December 2015. http://www.cdc.gov/nchs/data/nhis/earlyrelease/wireless201512.pdf.

Boreman, J. 2016. The Marine Recreational Information Program in the Mid-Atlantic Region. Presentation to NAS Committee. Washington, DC, February 25, 2016.

Boyle, J., F. Lewis, and B. Tefft. 2009. Cell phone mainly households: Coverage and reach for telephone surveys using RDD landline samples. *Survey Practice* 2(9).

Breidt, F. J. 2013. Recreational Fisheries Survey Methods. Statement before the Subcommittee on Fisheries, Wildlife, Oceans and Insular Affairs, 113th Congress, May 21, 2013. http://www.nationalacademies.org/OCGA/113Session1/testimonies/OCGA_145530.

Breidt, F. J., and J. R. Chromy. 2016. Marine Recreational Information Program Access Point Angler Intercept Survey. Support Statement Marine Recreational Information Program. OMB Control No. 0648-0659.

Breidt, F. J., L. Han-Lin, J. D. Opsomer, and D. A. Van Voorhees. 2012. A Report of the MRIP Sampling and Estimation Project: Improved Estimation Methods for the Access Point Intercept Survey Component of the Marine Recreational Fishery Statistics Survey. http://www.countmyfish.noaa.gov/projects/downloads/Final%20Report%20of%20New%20Estimation_Method_for_MRFSS_Data-01242012.pdf.

Brennan, K. 2016. Southeast Regional Headboat Survey. Presentation to NAS Committee. New Orleans, LA, May 27, 2016.

Brick, J. M., I. F. Cervantes, S. Lee, and G. Norman. 2011. Nonsampling errors in dual frame telephone surveys. *Survey Methodology* 37(1):1-12.

Carmichael, J., and D. Van Voorhees. 2014. MRIP Calibration Workshop II Final Report. North Charleston, SC. https://www.st.nmfs.noaa.gov/Assets/recreational/pdf/MRIPCalibrationWorkshopII_FinalReport.pdf.

Carmichael, J., and M. Duval. 2016. SAFMC Perspective NAS MRIP Review. Presentation to NAS Committee. Charleston, SC. April 16, 2016.

Chromy, J. R., S. M. Holland, and R. Webster. 2009. Consultants' Report: For-Hire Recreational Fisheries Surveys. Submitted to the For-Hire Work Group, National Marine Fisheries Service, March 2009. http://www.countmyfish.noaa.gov/projects/downloads/ ForHireReportFinal.pdf.

Cohen, A. H., and J. S. Krueger. 2016. Rising mercury, rising hostility: How heat affects survey response. *Field Methods* 28(2):133-152.

Coleman, A. 2004. The National Recreational Fishing Survey: The Northern Territory. NT Department of Business, Industry and Resource Development. Fisheries Group.

Cox, B. 1980. The weighted sequential hot deck imputation procedure. Pp. 721-726 in *Proceedings of the Survey Research Methods Section, American Statistical Association*. https://www.rti.org/publication/weighted-sequential-hot-deck-imputation-procedure.

Crabtree, R. 2016. MRIP Data Needs for Fishery Management Decisions. Presentation to NAS Committee. New Orleans, LA, May 27, 2016.

Curtin, R., S. Presser, and E. Singer. 2005. Changes in telephone survey nonresponse over the past quarter century. *Public Opinion Quarterly* 69(1):87-98.

Dahlhamer, J. M., M. Cynamon, J. Gentlman, A. Piani, and M. Weilier. 2010. Minimizing survey error through interviewer training: New procedures applied to the National Health Interview Study. *Section on Survey Research Methods* JSM 2010, American Statistical Association.

Dever, J., and R. Valliant. 2010. A comparison of variance estimators for poststratification to estimated control totals. *Survey Methodology* 36(1):45-56.

Dick, E. J. 2016. MRIP from a Stock Assessment Perspective. Presentation to NAS Committee. Irvine, CA, July 11, 2016.

Dillman, D., and B. L. Messer. 2010. Mixed-mode surveys. Pp. 551-574 in *Handbook of Survey Research*. P. V. Marsden and J. D. Wright, eds. Bingley, UK: Emerald Group Publishing Limited.

Dillman, D. A., J. D. Smyth, and L. M. Christian. 2014. *Internet, Phone, Mail, and Mixed-Mode Surveys: The Tailored Design Method.* 4th Edition. Hoboken, NJ: John Wiley and Sons, Inc.

Donaldson, D., G. Bray, B. Sauls, S. Freed, B. Cermak, P. Campbell, A. Best, K. Doyle, A. Strelcheck, and K. Brennan. 2013. For-Hire Electronic Logbook Pilot Study in the Gulf of Mexico. https://www.st.nmfs.noaa.gov/Assets/recreational/pdf/Charter_Boat_Logbook_Project_report.pdf.

FCC (U.S. Federal Communications Commission). 2016. Wireless Local Number Portability, May 18, 2016. https://www.fcc.gov/general/wireless-local-number-portability-wlnp.

Fisheries Statistics Division. 2016. Sampling and Estimation Methods for the New Fishing Effort Survey of the Marine Recreational Information Program. Documentation provided to NAS Committee. National Marine Fisheries Service, NOAA.

Fricker, S., and R. Tourangeau. 2010. Examining the relationship between nonresponse propensity and data quality in two national household surveys. *Public Opinion Quarterly* 74(5):934-955. https://poq.oxfordjournals.org/content/74/5/934.full.pdf+html.

Fuller, W. A. 2009. Some design properties of a rejective sampling procedure. *Biometrika* 96:933-944.

Gaskell, G., D. Wright, and C. O'Muircheartaigh. 2000. Telescoping of landmark events: Implications for survey research. *Public Opinion Quarterly* 64(1):77-89.

Goodman, L. A. 1960. On the exact variance of products. *Journal of the American Statistical Association* 55:708-713.

Groves, R. M. 1989. *Survey Errors and Survey Costs*. Hoboken, NJ: John Wiley and Sons, Inc.

Groves, R. M., and M. P. Couper. 1998. *Nonresponse in Household Interview Surveys*. New York: John Wiley and Sons, Inc.

Groves, R. M., E. Singer, and A. Corning. 2000. Leverage-Saliency Theory of Survey Participation: Description and Illustration. *Public Opinion Quarterly.* 60(3):299-308.

Groves, R. M., F. J. Fowler Jr., M. P. Couper, J. M. Lepkowski, E. Singer, and R. Tourangeau. 2009. *Survey Methodology*. 2nd Edition. Hoboken, NJ: John Wiley and Sons, Inc.

Haziza, D. 2009. Imputation and inference in the presence of missing data. *Handbook of Statistics* 29(Part A):215-246.

Haziza, D., and É. Lesage. 2016. A discussion of weighting procedures for unit nonresponse. *Journal of Official Statistics* 32(1):129-145.

Horrigan, J. B. 2015. The numbers behind the broadband 'homework gap.' *Factank: News in the Numbers,* April 20. http://www.pewresearch.org/fact-tank/2015/04/20/the-numbers-behind-the-broadband-homework-gap/.

Huttenlocher J., L. V. Hedges, and N. M. Bradburn. 1990. Reports of elapsed time: Bounding and rounding processes in estimation. *Journal of Experimental Psychology: Learning, Memory, and Cognition* 16(2):196-213.

Iannacchione, V. G. 2011. Research synthesis: The changing role of address-based sampling in survey research. *Public Opinion Quarterly* 75(3):556-575.

Ihde, T. F., M. J. Wilberg, D. A. Loewensteiner, D. H. Secor, and T. J. Miller. 2011. The increasing importance of marine recreational fishing in the US: Challenges for management. *Fisheries Research* 108:268-276.

Jones, C. M. 2001. Creel surveys. Pp. 442-449 in *Encyclopedia of Environmetrics*. A. H. El-Shaarawi and W. W. Piegorsch, eds. Hoboken, NJ: John Wiley and Sons, Inc.

Jones, C. M., and Pollock, K. H. 2013. Recreational angler survey methods: Estimation of effort, harvest, and released catch over a range of spatial and temporal scales. *Fisheries Techniques* 883-920.

Keeter, S., C. Kennedy, M. Dimock, J. Best, and P. Craighill. 2006. Gauging the impact of growing nonresponse on estimates from a national RDD telephone survey. *Public Opinion Quarterly* 70(5):759-779.

Keeter, S., K. McGeeney, C. Kennedy, and M. Schalk. 2015. Advances in telephone survey sampling: Balancing efficiency and coverage using several new approaches. Washington, DC: Pew Research Center. http://www.pewresearch.org/2015/11/18/advances-in-telephone-survey-sampling/.

Kelly, M. 2016. Technology and Innovation Trends in Electronic Reporting. Presentation to NAS Committee. July 11, 2016. Irvine, CA.

Kim, J. K., and J. K. Rao. 2009. A unified approach to linearization variance estimation from survey data after imputation for item nonresponse. *Biometrika* 96:917-932.

Kott, P. S. 2016. Calibration weighting in survey sampling. *WIREs Computational Statistics* 8(1):39-53. DOI: 10.1002/wics.1374.

Kott, P. S., and T. Chang. 2010. Using calibration weighting to adjust for nonignorable unit nonresponse. *Journal of the American Statistical Association* 105(491):1265-1275. DOI: 10.1198/jasa.2010.tm09016.

Kreuter, F. 2013. *Improving Surveys with Paradata: Analytic Uses of Process Information*. Hoboken, NJ: John Wiley and Sons, Inc.

Lavrakas, P. J. 2008. *Encyclopedia of Survey Research Methods*. Thousand Oaks, CA: SAGE Publications Ltd. DOI: 10.4135/9781412963947.

Link, M. W., M. P. Battaglia, M. R. Frankel, L. Osborn, and A. H. Mokdad. 2008. A comparison of address-based sampling (ABS) versus random-digit dialing (RDD) for general population surveys. *Public Opinion Quarterly* 72(1): 6-27.

Link, M. W., J. Murphy, M. F. Schober, T. D. Buskirk, J. H. Childs, and C. L. Tesfaye. 2014. Mobile Technologies for Conducting, Augmenting and Potentially Replacing Surveys Executive Summary of the AAPOR Task Force on Emerging Technologies in Public Opinion Research. *Public Opinion Quarterly* 78(4):779-787.

Liu, B., L. Stokes, G. Stunz, and T. Topping. 2016. Estimation of total from a population of unknown size and application to estimating recreational red snapper catch in Texas. *Biometrics* 77:1-24.

Lohr, S. L. 2010. *Sampling: Design and Analysis*. 2nd Edition. Boston, MA: Brooks/Cole Cengage Learning.

Lohr, S. L. 2007. Recent developments in multiple frame surveys. Pp. 3257-3264 in *Proceedings of the American Statistical Association*, Section on Survey Research Methods. https://www.amstat.org/sections/srms/proceedings/y2007/Files/JSM2007-000580.pdf.

Lohr, S. L., and J. K. Rao. 2000. Inference from dual frame surveys. *Journal of the American Statistical Association* 95(449):271-280.

Lucy, J., and A. Studholme, eds. 2002. *Catch and Release in Marine Recreational Fisheries*. Proceedings of the Symposium on Catch and Release in Marine Recreational Fisheries, Held at Virginia Beach, VA.

Martin, B., D. Chanda, B. Muffley, R. Allen, and T. Baum. 2016. New Jersey MRIP. Presentation to NAS Committee. Washington, DC, February 25, 2016.

McCarthy, N. 2015. The great decline of the landline. *Forbes Magazine*. http://www.forbes.com/sites/niallmccarthy/2015/02/27/the-great-decline-of-the-landline-infographic/#63a6a781d826.

Medway, R. L., and J. Fulton. 2012. When more gets you less: A meta-analysis of the effect of concurrent web options on mail survey response rates. *Public Opinion Quarterly* 76(4):733-746. DOI: 10.1093/poq/nfs047.

Mills, M. J. 1979. Alaska Statewide Sport Fish Harvest Studies. Alaska Department of Fish and Game. Federal Aid in Fish Restoration, Annual Performance Report, 1978-1979, Project F-9-11, 20 (SW-I-A).

Mills, M. J., G. L. Fidler, A. L. Howe, S. K. Sonnichsen, and K. W. Webster. 1986. Alaska Statewide Harvest Study Operational Plan. Juneau, AK: Alaska Department of Fish and Game, Division of Sport Fisheries. Unpublished document.

Morrison, W. E., and T. L. Scott. 2014. Review of Laws, Guidance, Technical Memorandums and Case Studies Related to Fisheries Allocation Decisions. NOAA Technical Memorandum NMFS-F/SPO-148. Silver Spring: U.S. Department of Commerce, National Oceanic and Atmospheric Adminstration. http://www.nmfs.noaa.gov/sfa/management/allocation/morrison_scott_allocation_report.pdf.

Mumford, D. 2016. North Carolina Department of Environmental Quality, Division of Marine Fisheries. Presentation to NAS Committtee. Washington, DC, February 25, 2016.

NMFS (National Marine Fisheries Service). 2012a. MRIP Road Show: Florida Trip Summary and Next Steps. Unpublished NMFS Document (Draft). May 24, 2012.

NMFS. 2012b. MRIP Road Show: New Jersey Trip Summary and Next Steps. Unpublished NMFS Document (Draft). April 18, 2012.

NMFS. 2013. Supporting statement for Marine Recreational Information Program Fishing Effort Survey, OMB Control No. 0648-0652. http://reginfo.gov/public/do/PRAViewDocument?ref_nbr=201306-0648-005.

NMFS. 2014a. MRIP Data User Handbook, December 2014. http://www.st.nmfs.noaa.gov/recreational-fisheries/MRIP-Handbook/MRIP_ handbook.pdf.

NMFS. 2014b. Statement of Work. For Hire Survey and Large Pelagics Telephone Survey 2014-2018.

NMFS. 2015. Fisheries of the United States, 2014. NOAA Current Fisheries Statistics No. 2014. http://www.st.nmfs.noaa.gov/Assets/commercial/fus/fus14/documents/FUS2014.pdf.

NMFS. 2016a. Marine Recreational Information Program Fishing Effort Survey Transition Progress Report, October 28, 2016. http://www.st.nmfs.noaa.gov/Assets/recreational/pdf/2015_FES_Progress_Report.pdf.

NMFS. 2016b. MRIP 2016 Strategic Communications Plan.

NOAA (National Oceanic and Atmospheric Administration). 2009. *National Saltwater Angler Registry Opens on New Year's Day*. http://www.countmyfish.noaa.gov/newsroom/downloads/RecfishingregistryPR_09.pdf.

NOAA. 2014. Recreational Fishing Catch and Effort Data Collection. http://www.nmfs.noaa.gov/sfa/management/recreational/documents/2014_summit_surveys_overview_final.pdf.

NOAA. 2016a. Marine Recreational Information Program 2015-2016 Implementation Plan Update. Report by the NOAA Fisheries' Office of Science and Technology, https://www.st.nmfs.noaa.gov/Assets/recreational/pdf/FINAL-updated-implementation-plan-3.22.16.pdf.

NOAA. 2016b. Access Point Angler Intercept Survey. http://www.st.nmfs.noaa.gov/st1/recreational/Intercept_survey.htm.

NRC (National Research Council). 1999. *Sustaining Marine Fisheries*. Washington, DC: National Academy Press.

NRC. 2000. *Improving the Collection, Management, and Use of Marine Fisheries Data*. Washington, DC: National Academy Press. https://doi.org/10.17226/9969.

NRC. 2006. *Review of Recreational Fisheries Survey Methods*. Washington, DC: The National Academies Press.

OMB (U.S. Office of Management and Budget). 2006. Standards and Guidelines for Statistical Surveys, September 2006. https://www.whitehouse.gov/sites/default/files/omb/inforeg/statpolicy/standards_stat_surveys.pdf.

PFMC (Pacific Fishery Management Council). 2016. Pacific Coast Groundfish Fishery Management Plan. Portland, OR: PFMC.

Pollock, K. H., C. M. Jones, and T. Brown. 1994. *Angler Survey Methods and Their Application in Fisheries Management*. American Fisheries Society Special Publication 25. Bethesda, MD: American Fisheries Society.

Potter, F. J. 1990. Methods for Extreme Weights in Sample Surveys (Dissertation). The University of North Carolina at Chapel Hill.

Quinn, T. J., and R. B. Deriso. 1999. *Quantitative Fish Dynamics*. New York: Oxford University Press.

Reist, B. 2012. 2010. Census Investigation of Methods to Evaluate the Coverage of Group Quarters Population. 2010 Census Planning Memoranda Series, No. 240, September 27, 2012. Washington, DC: US Census Bureau.

Sakshaug, J. W., T. Yan, and R. Tourangeau. 2010. Nonresponse error, measurement error, and mode of data collection tradeoffs in a multi-mode survey of sensitive and non-sensitive items. *Public Opinion Quarterly* 74(5):907-933.

Särndal, C. E., and S. Lundström. 2005. *Estimation in Surveys with Nonresponse*. Hoboken, NJ: John Wiley and Sons, Inc.

Särndal, C. E., B. Swensson, and J. Wretman. 1992. *Model Assisted Survey Sampling*. New York: Springer-Verlag.

Sauls, B., N. Bartlett, M. Burton, S. Crooke, B. Clifford, J. Didden, M. Kasprzak, S. Meyer, G. Shepherd, T. Sminkey, M. Sosik, B. Zales, J. Chromy, S. Holland, and R. Webster. 2008. Inventory of For-Hire Data Collections in the United States and U.S. Territories. https://www.st.nmfs.noaa.gov/mdms/public/finalReport.jsp?ReportID=352 .

Scholtus, S. 2011. *Correction of Rounding, Typing and Sign Errors with the Deducorrect Package*. The Hague/Heerlen: Statistics Netherlands.

SEDAR (Southeast Data Assessment and Review). 2013. SEDAR 31: Gulf of Mexico Red Snapper. Stock Assessment Report. North Charleston, SC: SEDAR.

Shao, J., and P. Steel. 1999. Variance estimation for survey data with composite imputation and nonnegligible sampling fractions. *Journal of the American Statistical Association* 94:254-265.

Sharpe, L. 2016. Updated MRIP plan highlights strategic approach, regional implementation, and partner collaboration. March 23, 2016. https://www.st.nmfs.noaa.gov/Assets/recreational/pdf/newscastUpdatedImplementationPlan-3.23.16.pdf.

Sharpe, L., and D. Bard. 2016. MRIP Communications. Presentation to NAS Committee. Irvine, CA, July 12, 2016.

Shook-Sa, B. E., D. B. Currivan, J. P. McMichael, and V. G. Iannacchione. 2013. Extending the coverage of address-based sampling frames: Beyond the USPS computerized delivery sequence file. *Public Opinion Quarterly* 77(4):994-1005.

Stunz, G., D. Yoskowitz, and M. Robillard. 2014. iSnapper: An "App" for Collecting Catch Data. Presentation given at the Allocation Workshop in Seattle, WA, September 2014. https://www.st.nmfs.noaa.gov/Assets/economics/fisheries/commercial/allocation-workshop/yoskowitziSnapper AllocationWorkshopSeattleSept2014.pdf.

Sylvia, A., M. J. Wilberg, J. Wiedenmann, and T. J. Miller. 2016. Effects of assessment interval and data management lag on fishery management performance: Strategies for improvement (Master's Thesis). Digital Repository at the University of Maryland.

USFWS (U.S. Fish and Wildlife Service). 2016. Diary Surveys: Migratory Bird Hunter Survey. https://www.fws.gov/birds/surveys-and-data/harvest-surveys/diary-surveys.php/.

Valliant, R. L. 2004. The effect of multiple weight adjustments on variance estimation. *Journal of Official Statistics* 20(1):1-18.

Valliant, R., J. M. Brick, and J. A. Dever. 2008. Weight adjustments for the grouped jackknife variance estimator. *Journal of Official Statistics* 24(3):469-488.

Valliant, R., J. A. Dever, and F. Kreuter. 2013. *Practical Tools for Designing and Weighting Sample Surveys*. New York: Springer.

Wallace, F., K. Williams, R. Towler, and K. McGauley. 2015. Innovative Camera Applications for Electronic Monitoring. Pp. 105-117 in *Fisheries Bycatch: Global Issues and Creative Solutions*. G. H. Kruse, H. C. An, J. DiCosimo, C. A. Eischens, G. S. Gislason, D. N. McBride, C. S. Rose, and C. E. Siddon, eds. Fairbanks, AK: Alaska Sea Grant, University of Alaska Fairbanks.

Wolter, K. 2007. *Introduction to Variance Estimation*, 2nd Edition. New York: Springer-Verlag.

Appendix A

Committee and Staff Biographies

COMMITTEE

Luiz Barbieri (*Co-chair*) is the Science and Research Director for the Marine Fisheries Research Program at Florida's Fish and Wildlife Research Institute, based out of St. Petersburg, Florida. He has an extensive background in marine fisheries science and policy and serves as a key representative on several scientific advisory panels and committees including the Atlantic States Marine Fisheries Commission, the Scientific and Statistical Committee (SSC) for both the Gulf of Mexico and South Atlantic Fishery Management Councils, and as a senior advisor to the Fisheries Leadership and Sustainability Forum, Nicholas Institute for Environmental Policy Solutions at Duke University. Dr. Barbieri received a B.S. in biology from Santa Ursula University, an M.S. in biological oceanography from Rio Grande University, and a Ph.D. in marine fisheries science from the College of William and Mary.

Cynthia M. Jones (*Co-chair*) is a professor and eminent scholar in the Ocean, Earth, and Atmospheric Sciences department of Old Dominion University. Her research interests include demography of fish based on age evaluation, stock assessment, recreational angler surveys, simulation modeling, and quantitative statistics. Dr. Jones has been the recipient of multiple awards and honors, including AAAS Fellow, Outstanding Virginia Scientist, Outstanding Professor, and Fulbright Senior Scholar Award. Dr. Jones served on the National Research Council (NRC) Ocean Studies Board from 2005 to 2007. She was also a member of multiple previous NRC studies, including the Committee for Review of the National Marine Fisheries Service: Use of Science and Data in Management and Litigation, which she chaired. Dr. Jones received her B.A. in zoology from

Boston University. She received her M.S. and Ph.D. in oceanography from the University of Rhode Island.

Jill A. Dever is Senior Survey Statistician at RTI International in Washington, DC. Her current research interests include statistical and methodological issues for dual-frame random-digit-dial and address-based sample surveys, along with nonprobability studies and survey weight calibration with estimated population value. In addition, her experience includes creating software for optimizing complex sample designs; constructing linearization and replicating analysis weights using calibration techniques; and analyzing data from complex surveys. To date, she has worked on a variety of surveys that address pertinent issues of our times in the areas of health care, education, and the U.S. military. Dr. Dever received her B.S. in mathematics from the University of Louisville, located in her hometown. She received her M.S. in biostatistics from the University of North Carolina Chapel Hill and her Ph.D. in survey methodology from the Joint Program in Survey Methodology at the University of Maryland College Park.

David Haziza is a professor in the Department of Mathematics and Statistics at the Université de Montréal. His research interests include theory and application of survey sampling, and specifically, inference in the presence of missing data or influential units, resampling methods, and calibration. Dr. Haziza received his Ph.D. in statistics from Carleton University.

Jeffrey C. Johnson is a professor of anthropology at the University of Florida. His research interests include network models of complex biological systems and the application of continuous time Markov chain and exponential random graph models to the study of trophic dynamics in food webs. His most recent work involves the development of methods for the reliable tagging, coding and network modeling of large corpora of related texts. He has published extensively in anthropological, sociological, biological, and marine science journals and was the founding editor of the *Journal of Quantitative Anthropology* and co-editor of *Human Organization*. He is currently an associate editor for the *Journal of Social Structure* and *Social Networks*. Dr. Johnson received his Ph.D. in anthropology from the University of California, Irvine.

Bruce M. Leaman has been the Executive Director of the International Pacific Halibut Commission since 1997. His research interests have included fisheries management, stock assessment, reproductive biology, and population dynamics. Dr. Leaman served on the 2006 National Research Council Committee on Review of Recreational Fisheries Survey Methods. Dr. Leaman received his B.Sc. from Simon Fraser University. He also received his M.Sc. and Ph.D. from the University of British Columbia.

Thomas J. Miller is a professor of fisheries and bioenergetics and population dynamics, as well as the Director of the Chesapeake Biological Laboratory at the University of Maryland's Center for Environmental Science. Dr. Miller's research interests include recruitment and population dynamics of aquatic animals, fish early life history, and blue crabs. His relevant National Research Council service includes membership on the Committee on Sustainable Water and Environmental Management in the California Bay-Delta and the Panel to Review California Draft Bay Delta Conservation Plan. He is also currently serving as a member of the Panel on Life Sciences. Dr. Miller received his B.Sc. in human and environmental biology at the University of York. He later received his M.S. at North Carolina State University in ecology and his Ph.D. in zoology, also from North Carolina State University.

Sean P. Powers is a professor and chair of marine sciences at the University of South Alabama and senior marine scientist at the Dauphin Island Sea Lab. His research focuses on the ecology of coastal/estuarine fishes and benthic invertebrates, particularly those that support commercial and recreational fisheries. The goal of his research program is to provide scientifically sound information to direct conservation and restoration efforts of marine fisheries and the habitats that these species rely upon. He has served on one National Research Council committee, the Committee on U.S. Army Corps of Engineers Water Resources, Science, Engineering, and Planning. Dr. Powers received his Ph.D. in zoology from Texas A&M University.

Steve Williams is a Senior Program Manager for the Pacific States Marine Fisheries Commission. Mr. Williams is primarily responsible for program oversight of the Pacific RecFIN database program as well as several other Columbia River and Pacific coast marine projects dealing with salmon and highly migratory species management issues. Mr. Williams spent 39 years as both a fish and a wildlife manager for the Oregon Department of Fish and Wildlife. His management experience covers a broad range from field sampling of both marine and freshwater fisheries to a period as Deputy Director of the Oregon Department of Fish and Wildlife. He served as the Oregon state representative to the Pacific Fishery Management Council for approximately 5 years. Mr. Williams received his B.S. in fisheries from Oregon State.

STAFF

Susan Roberts became the director of the Ocean Studies Board in April 2004. Dr. Roberts received her Ph.D. in marine biology from the Scripps Institution of Oceanography. Prior to her position at the Ocean Studies Board, she worked as a postdoctoral researcher at the University of California, Berkeley, and as a senior staff fellow at the National Institutes of Health. Dr. Roberts' research experience has included fish physiology and biochemistry, marine bacterial symbioses, developmental cell biology, and environmentally induced leukemia. Dr. Roberts specializes in the science and management of living marine resources. She has served as study director for reports produced by the National Research Council on topics covering a broad range of ocean science, marine resource management, and science policy issues. She is a member of the U.S. National Committee for the Intergovernmental Oceanographic Commission (IOC) and serves on the IOC panel for the Global Ocean Science Report. Dr. Roberts is a member of AAAS, American Geophysical Union, and the Association for the Sciences of Limnology and Oceanography. She is an elected Fellow of the Washington Academy of Sciences.

Stacee Karras is a program officer with the Ocean Studies Board. She joined the National Academies of Science, Engineering, and Medicine in 2012 as a fellow, and served as a research associate for the Ocean Studies Board between 2013 and 2015. She then served as an associate program officer until 2016, when she took on her current role. She received her B.A. in marine affairs and policy with concentrations in biology and political science from the University of Miami in 2007. The following year she received an M.A. in marine affairs and policy from the University of Miami's Rosenstiel School of Marine and Atmospheric Science. In 2012, she earned her J.D. from the University of Virginia, School of Law.

Michael Cohen is a senior program officer for the Committee on National Statistics. He has led or served as contributing staff on a wide range of studies on the U.S. census and the modeling and reliability of defense systems. He also serves as a consultant on statistical analysis for other divisions in the National Academies of Sciences, Engineering, and Medicine. Previously, he was a mathematical statistician at the Energy Information Administration and held positions at the School of Public Affairs at the University of Maryland and at Princeton University. His general area of interest is the use of statistics in public policy, with a particular focus on census undercount, model validation, and robust estimation. He is a fellow of the American Statistical Association and an elected member of the International Statistical Institute. He has a B.S. in mathematics from the University of Michigan and an M.S. and a Ph.D. in statistics from Stanford University.

David Policansky received his Ph.D. in biology from the University of Oregon, where he studied evolutionary biology and ecology. He has published on life-history transitions, including the cost and timing of sexual reproduction in plants and animals; he also has published on fisheries and the interface between science and policy and on the inheritance of asymmetries in flounders. In his more than 30 years at the National Academies of Science, Engineering, and Medicine he has been involved in more than 35 reports, many as project director. His work has focused on management of natural resources, natural restoration, information for environmental decision making, reviews of large federal programs, and on endangered species, among other topics.

Payton Kulina (until September 2016) joined the Ocean Studies Board in June 2013 as a senior program assistant. He graduated from Dickinson College in 2010 receiving a B.A. in policy management. He is currently pursuing an M.S. degree in finance through the Kogod School of Business at American University. Prior to this position, Mr. Kulina worked as a coordinator with BP Alternative Energy, also in Washington, DC.

Allie Phillips (starting September 2016) graduated in May 2016 from Colby College in Waterville, Maine, where she received a B.A. in environmental studies: policy. As an undergraduate, she held internships at the Environmental League of Massachusetts and the New England Aquarium. She joined the Ocean Studies Board as a program assistant in September 2016.

Appendix B

Review of Recreational Fisheries Survey Methods (NRC, 2006) Summary

INTRODUCTION

Recreational fishing in the United States is an important social and economic component of many marine fisheries. However, in some cases, recreational fishing takes more fish than commercial fishing, and in an increasing number of cases, recreational fishing is the main source of fishing mortality. In addition, current assessments indicate that some marine recreational fisheries have exceeded their quotas, raising concern because fishing effort in marine recreational fisheries is projected to increase. It is important that catch monitoring systems are adequate for timely management of these fisheries.

Marine recreational fisheries are not monitored with the same rigor as commercial fisheries. However, as concerns about the effects of all types of fishing have grown, more attention has been paid to the possible impacts of marine recreational fishing. The growing interest in the effects of recreational fishing on fish stock size and composition has led to increased demands for timely and accurate data. Although the National Marine Fisheries Service (NMFS) of the National Oceanic and Atmospheric Administration implemented the Marine Recreational Fisheries Statistics Survey (MRFSS) in 1979 to obtain statistics about marine recreational fisheries, management goals and objectives have changed since then, as has the complexity of the recreational fishing sector. The need for and use of marine recreational fishery statistics in science and management have changed as well. This committee has identified several areas in which designers of sampling programs, data collectors, and users of recreational fisheries data appear to have incomplete communication, mismatched criteria, or other obstacles.

The MRFSS has two major components: an onsite component, in which anglers are intercepted and interviewed on the water or at sites such as marinas where they access the water, and an offsite component, in which anglers are contacted and surveyed by telephone after their trips are completed. There has been widespread criticism of the nature and use of the MRFSS information. The MRFSS was (and is) intended to be a national program, but not all coastal states participate. In some cases, states have their own surveys of recreational fish landings instead of the MRFSS; in other cases, states have surveys that complement the MRFSS. In addition to this lack of uniformity of coverage, the quality of the MRFSS data for management purposes has also been questioned.

Indeed, it is much more difficult to collect data on recreational saltwater anglers than on commercial fishing operations. There are far more saltwater anglers than commercial fishermen—approximately 14 million anglers fished annually in recent years—and they do not land their catches at specific points where there are dealers, as do commercial fishermen. In addition, there are many modes of fishing (e.g., anglers who fish from head boats or charter boats, with guides,[1] from shore, on private boats, from private property), and many anglers release fish they catch. Some anglers travel far to fish and often fish only a few times each year, which makes them difficult to encounter in surveys. Others, who live within 50 miles of the coast, are much more likely to be intercepted by the MRFSS. Finally, most surveys of anglers depend to some degree on the anglers' recall and willingness to volunteer valid information. As a result, designing a survey that will provide accurate and timely information, with good coverage and at acceptable cost, is a major challenge.

Despite the complexity of the challenge and its importance for fishery management, the MRFSS staff have been severely handicapped in their efforts to implement, operate, and improve the MRFSS, including implementing the recommendations of earlier reviews. It is not reasonable to expect such a small staff—and one that lacks a Ph.D.-level mathematical statistician—to operate a national survey of such complexity, despite the dedication of the small staff the MRFSS does have.

In addition, the MRFSS is severely limited by the lack of a universal sampling frame for all saltwater anglers, a lack that is not of the MRFSS's own making. To make matters even more difficult, some of the data that the MRFSS depends on are collected by states, which use a variety of data-collection and sampling protocols. Finally, the financial resources allocated to the MRFSS are modest in comparison to the challenge. This committee's findings and recommendations should be viewed with this in mind.

---

[1] Head boats, also called party boats, take large groups of anglers (sometimes as many as 100) on fishing trips; the groups usually are not preformed. Charter boats (also occasionally called party boats) take smaller groups of anglers, usually four to eight, most often in preformed groups. Guided trips are trips in which a guide takes one or two anglers in a smaller boat. These different categories operate under different U.S. Coast Guard and state license requirements. Throughout this report, these sectors are collectively referred to as the for-hire sector.

THE PRESENT STUDY

To help identify solutions to some of the above problems, NMFS asked the National Academies to assemble a committee to review current marine recreational fishing surveys and to make recommendations for improvements—especially to the MRFSS—and to recommend the implementation of possible alternative approaches. (See Box B.1 for the committee's statement of task.)

In response, the National Research Council (NRC) of the National Academies established the Committee on the Review of Recreational Fishing Survey Methods, composed of experts in survey design and statistics, biological statistics, fishery management, and the economics and sociology of recreational

BOX B.1
Statement of Task

This study will critically review the types of survey methods used to estimate catch per unit effort and effort in recreational fisheries, including state and federal cooperative programs. The committee will examine representative survey types but will not evaluate every regional or state survey method currently in use. The study will consider the match or mismatch between options for collecting recreational fisheries data and alternative approaches for managing recreational fisheries.

In particular, the committee will assess current types of survey methods giving consideration to

- The suitability for monitoring different types of fishing (e.g., charter boats versus private boats, offshore versus near-shore species, fisheries with temporally or spatially restricted fishing seasons).
- The adequacy for providing the quality of information needed to support various approaches for managing recreational fisheries, with reference to how the management approach might be restricted by the type of survey method, stratification scheme, and sample size required. For example, is the management time frame (in-season, annual, or multiyear) consistent with temporal design of the survey? Is the geographic scale of management (e.g., state versus regional) appropriate for the resolution provided by the survey? How would the survey design need to be modified to match the requirements of the management approach?
- Make recommendations regarding possible improvements to current surveys and/or possible implementation of alternative approaches, including setting priorities for revising monitoring methods that will yield the greatest improvements in effort and catch per unit effort estimates.

Current survey methods and recommended alternatives will be compared with relation to costs, sources of bias, precision, and timeliness.

fishing. The background and support for the conclusions and recommendations presented below are found in subsequent chapters.

CONCLUSIONS AND RECOMMENDATIONS

General Conclusions

- The committee agrees with conclusions of previous NRC committees that marine recreational fishing is a significant source of fishing mortality for many marine species and that adequate scientific information on the nature of that mortality in time and space is required for successful management of those species.
- Marine fisheries management goals, objectives, and context have changed since the MRFSS was begun in 1979. Management decisions are often made at finer spatial and temporal scales than they were earlier, the mix of recreational and commercial fishing has changed for many areas and species, and stock-assessment models now make greater use of data from recreational fisheries.
- The MRFSS is in need of additional financial resources so that technical and practical expertise can be added to assist in a major overhaul of the design, implementation, and analysis of data from the MRFSS. Both the telephone and access components of the current approach have serious flaws in design or implementation and use inadequate analysis methods that need to be addressed immediately. This committee's review has focused primarily on the MRFSS, but many of the component surveys of the MRFSS conducted by state agencies (with various degrees of federal funding) suffer from the same shortcomings as does the central MRFSS. As a result, many of this committee's recommendations apply to state surveys as well as to the MRFSS.
- Many of the independent surveys conducted by the states, as well as state-run surveys that are components of the MRFSS, are different from each other and from the central MRFSS in important ways, including sampling, data collection, and preparation of estimators.
- The committee concludes that users' concerns about the use of the MRFSS in fishery management are justified by the above-mentioned weaknesses, but they also result from inadequate communication and outreach on the part of the MRFSS managers at NMFS.
- The for-hire sector of marine recreational fisheries (i.e., charter, guide, and head boat operations) is more like a commercial sector than it is like the private-angler sector.

General Recommendations

- The MRFSS (as well as many of its component or companion surveys conducted either indirectly or independently) should be completely redesigned to improve its effectiveness and appropriateness of sampling and estimation procedures, its applicability to various kinds of management decisions, and its usefulness for social and economic analyses. After the revision is complete, provision should be made for ongoing technical evaluation and modification, as needed, to meet emerging management needs. To improve the MRFSS, the committee further recommends that the existing MRFSS program be given a firm deadline linked to sufficient program funding for implementation of this report's recommendations.
- A much greater degree of standardization among state surveys, and between state surveys and the central MRFSS, should be achieved. This will require a much greater degree of cooperation and coordination among the managers of the various surveys. The for-hire sector of marine recreational fisheries should be considered a commercial sector, and survey methods and reporting requirements for that sector therefore should be different from those for private anglers.

Sampling Issues Conclusions

- The committee concludes that the current methods used in the MRFSS for sampling the universe of anglers and for determining their catch and effort are inadequate. Sampling of each group of anglers (i.e., private, guided, head boat, charter boat) presents challenges that can differ across the groups. Two complementary methods of sampling are used in the MRFSS. One is onsite (i.e., intercepting anglers while they are fishing or at their access [landing] points). The other is offsite, which includes a variety of sampling techniques for contacting anglers after they have completed their trips. Both onsite and offsite methods suffer from weaknesses that may lead to biases in catch and effort estimation. Finally, the estimation procedure for information gathered onsite does not use the nominal or actual selection probabilities of the sample design and therefore has the potential to produce biased estimates for both the parameters of interest and their variances.
- Onsite methods fail to intercept anglers who have private access to fishing waters or intercept them only sporadically. It is impossible, using current methods, to obtain information on the target species of anglers who have private access. In addition, various physical, financial, and operational constraints often lead to spatial or temporal biases in onsite sampling coverage that are not adequately accounted for in the estimation equations.

- Offsite sampling methods that rely on telephone interviews are complicated by the increasing use of cellular telephones, especially in surveys of residents of coastal counties. This is because cellular telephones are not restricted to a geographic region as are landline telephones. If cellular telephones are excluded, then undercoverage of the survey will be increasingly problematic over time as the number of people who use only cellular telephones is growing. The existing random-digit-dialing (RDD) survey suffers in efficiency from the low proportion of fishing households among the general population and may allow bias in estimation from its restriction to coastal counties only.
- The existing RDD survey suffers in efficiency from the low proportion of fishing households among the general population and may allow bias in estimation from its restriction to coastal counties only.
- Reliance on fishing license–based lists of saltwater anglers is not yet feasible as a means of improving offsite sampling methods to avoid the inefficiency of RDD, undercoverage due to cellular telephone use, and restriction to coastal counties. Although many states collect angler information when a saltwater fishing license is purchased, there are license exemptions based on age, residence, access points, existence of a boat license, mode of fishing, and other factors. As a result, angler information for those states is incomplete. Some states have more complete information than others, and in the states that have no saltwater license, there is no list of saltwater anglers. The lack of a universal sampling frame (registry or license requirement) for all saltwater anglers is a major impediment to the development of a reliable and accurate survey program.
- Catch and release fishing (release of fish that survive capture) is increasingly common in many marine recreational fisheries. Although some fish survive capture and release, mortality may be high, in some cases exceeding 50 percent. The survey fails to provide a valid and reliable method of adequately accounting for fish caught and *not* brought to the dock (including fish released alive or dead, as well as fish caught for bait or given away before reaching the dock). This shortcoming affects estimates of catch and total removals.
- The correct identification of fish species, especially in places with diverse fish faunas, is a difficult challenge, both for many anglers and for those conducting surveys. Incorrect identification obviously has the potential to lead to incorrect conclusions from survey data.

Sampling Issues Recommendations

- A comprehensive, universal sampling frame with national coverage should be established. The most effective way to achieve this is through a national registration of all saltwater anglers or through new or existing state saltwater license programs that would allow no exemptions[2] and that would provide appropriate contact information from anglers fishing in all marine waters, both state and federal. Any gaps in such a program (e.g., a lack of registration in a particular region or mode, exemptions of various classes of anglers) would compromise the use of the sampling frame and, hence, the quality of the survey program. An updated, complete registration list would greatly improve sampling efficiency in terms of time and cost. Although these savings might not cover the entire cost of maintaining such a database, the benefit from the increased quantity and quality of the data would be worth the extra cost, especially if there is an associated increase in public confidence in the final estimates.

- Future telephone surveys should be based on the above universal sampling frame.
- Charter boat, head boat, and other for-hire recreational fishing operations should be required to maintain logbooks of fish landed and kept, as well as fish caught and released. Providing the information should be mandatory for continued operation in this sector, and all the information should be verifiable and made available to the survey program in a timely manner.
- Additional studies are needed to understand the extent to which fish are kept and inspected, as well as the extent of catch not available for inspection to improve the accuracy of catch estimates.
- Panel surveys, which contact individual anglers repeatedly over time, should be considered in recreational fishing surveys to gather angler trend data and to improve the efficiency of data collection.
- The onsite sampling frame for the MRFSS should be redesigned. The estimation procedure critically depends on the assumption that catch rate does not vary according to the nature of the access point. In particular, small or private access points that most likely are missed might have different catch rates than larger access points, which would lead to bias in the resulting estimators. In addition, the sampling process requires greater quality control (less latitude on the part of the samplers) than it has at present. (See the recommendation below for the establishment of an independent research group to investigate matters such as these.)
- Dual-frame procedures should be used wherever possible to reduce sample bias. For example, if a state has an incomplete list frame based on licenses, the use of an additional sampling frame of the state's resi-

[2] There is no scientific reason that a state should not continue to allow certain groups (e.g., seniors) to fish for free, as long as everyone is required to register in the universal sampling frame or have a state saltwater license.

dents (e.g., RDD) would reduce the bias. The existence of a universal frame described above would make this approach unnecessary for offsite sampling.

- Internet surveys should be considered for their potential use in recreational fishing surveys, especially in panel surveys, as a way for anglers to submit information.

Statistical Estimation Issues Conclusions

- The designs, sampling strategies, and collection methods of recreational fishing surveys do not provide adequate data for management and policy decisions. Unknown biases in the estimators from these surveys arise from reliance on unverified assumptions. Unless these assumptions are tested and the degree and direction of bias reliably estimated, the extent to which the biases affect final estimates will remain unknown.
- The statistical properties associated with data collected through different survey techniques differ and often are unknown. The current estimators of error associated with various survey products are likely to be biased and too low. It is necessary, at a minimum, to determine how those differences affect survey results that use differing methods.

Current analysis procedures used in the MRFSS do not exploit the current knowledge of finite population sampling theory. The current estimates are particularly deficient when applied to small areas because they do not use information in adjoining areas or time periods, nor do they consider relationships between species that occur together. Therefore, they are of lower precision than would be possible if this information were used. Improvements in these estimates would be of great use to managers who need to make quick decisions concerning spatial areas that are smaller than typical in the early years of the MRFSS.

Statistical Estimation Issues Recommendations

- The statistical properties of various sampling, data-collection, and data-analysis methods should be determined. Assumptions should be examined and verified so that biases can be properly evaluated.
- A research group of statisticians should design new analyses based on current developments in sampling theory. These examinations should include experimentation, such as specific sampling of activities like nighttime fishing or fishing from private property, whose current under-representation in the MRFSS sampling has the potential to create bias.

Human Dimensions Conclusions

- The MRFSS was not designed with human dimensions data (i.e., collection of social, behavioral, attitudinal, and economic data) in mind. The qualities of social, economic, and other human dimensions data have been compromised for many of the same reasons that the biological data have been compromised, including such issues as those related to coastal populations, telephone surveys, and sampling protocol. The human dimensions data have been further compromised by simply being added onto the biological data-collection efforts that have different sampling requirements and survey design needs. Current surveys are largely focused on biological factors (e.g., numbers, sizes, and species of fish landed) and not on human dimensions factors. The statistical and sampling problems associated with social, behavioral, attitudinal, and economic data often can be considerably different from those associated with biological factors.

  If the number of marine fishing trips increases, it is likely that additional fishing access sites will be developed. In addition, social and environmental changes (e.g., changes in the distribution and numbers of people, a major hurricane) also can affect the availability and use of access sites. To ensure adequate coverage of the recreational fishery, a periodic updating of lists and descriptions of fishing locations and access sites is needed.

Human Dimensions Recommendations

- An independent national trip and expenditure survey should be developed to support economic valuation studies, impact analyses, and other social and attitudinal studies. The sampling and survey procedures of the independent survey should be designed for the purpose of social and economic, not biological, analyses.
- Add-on surveys for human dimensions should be continued but in a more focused way than currently is done to target specific management needs and to supplement the national data as needed.
- The national database on marine recreational fishing sites and their characteristics should be enhanced to support social, economic, and other human dimensions analyses. Sites should be defined at levels as fine as possible. The data set should include site characteristics that matter to anglers in making fishing choices, such as boat ramps, facilities, natural amenities, parking, size, and type (e.g., beach, pier, launch point). To account for changes in the number and patterns of trips and the changing characteristics of sites, a periodic updating of the data should be conducted.

Program Management and Support Conclusions

- A large number of complex technical issues associated with surveys of marine recreational fishing remain unsolved, and a significant investment in intellectual and technical expertise is needed.
- A greater degree of coordination between federal, state, and other survey programs is necessary to achieve the national perspective on marine recreational fisheries that is needed.
- The recommended changes to the design and operation of the MRFSS and its continued development and operation will require additional funding above current levels.

Program Management and Support Recommendations

- A permanent and independent research group should be established and funded to continuously evaluate the statistical design and adequacy of recreational fishery surveys and to guide necessary modifications or new initiatives. Human dimensions expertise should be included as well.
- Additional funding is needed for a survey office devoted to the management and implementation of marine recreational surveys, including coordination between surveys conducted in various state and federal agencies.

Communication and Outreach Conclusions

- It is difficult for individual anglers to see the effects of recreational fishing on their target species and to distinguish daily and seasonal fluctuations from trends. As a result, no matter how well designed and implemented a marine recreational survey is, it will not fully succeed without the cooperation of anglers. Unless anglers believe that the survey is well designed and implemented and that it is being used intelligently to address appropriate management issues, they are unlikely to participate.
- In particular, anglers need to have a basic understanding of the relationship between a statistically based sampling scheme and the frequency with which each of them is (or is not) contacted by a data collector.
- If anglers believe that their input is influencing the design and use of surveys, they are more likely to be satisfied with those surveys than otherwise.
- If anglers understand the basic purposes and decisions to which recreational fishing survey data are being applied and how those data are interpreted and used, they are more likely to feel confident that the ap-

proaches used are legitimate and are more likely to participate willingly and provide valid information.

Communication and Outreach Recommendations

- Outreach and communication should be improved in several ways. The MRFSS managers should advise anglers and data users on the constraints that apply to the use of the data for various purposes. Managers and anglers also should be informed clearly about any limitations of the data.
- Outreach and communication should be institutionalized as part of an ongoing MRFSS program so their importance is acknowledged and appropriate expertise can be developed.
- Angler associations should be engaged as partners with survey managers through workshops, data collection, survey design, and participation in survey advisory groups. Many NRC and other reports stress the importance of using local and traditional knowledge, capacity building, and local communities in knowledge-gathering and dissemination activities. These recommendations apply, as well, to the recreational fishing community.

Appendix C

Table of National Research Council (2006) Recommendations

TABLE C.1 2006 National Research Council recommendations and Ranking of NMFS responses. The first column is a list of recommendations from the 2006 NAS review, second column lists the primary chapter(s) in this report that discuss the matters raised in those recommendations (recognizing that some topics are discussed in multiple chapters), and the third column is a list of this committee's general evaluations of MRIP's responses to the 2006 report. Five "+"s indicates that the response has been comprehensive and has addressed the major components of the recommendation. Between two and four "+"s indicates some progress but NMFS had not fully addressed the recommendation. This may include consideration of difficult technical issues not yet solved, as well as partial or incomplete responses from MRIP. A single "+" means that there has not been substantial progress for various reasons. In circumstances where technological advances, new approaches, or new information have decreased the applicability or relevance of the 2006 recommendation, "N/A" is used.

2006 Recommendations	Relevant Chapter(s) in Current Report	Ranking
GENERAL		
1. The MRFSS (as well as many of its component or companion surveys conducted either indirectly or independently) should be completely redesigned to improve the effectiveness and appropriateness of sampling and estimation procedures, applicability to various kinds of management decisions, and usefulness for social and economic analyses. After the revision is complete, provision should be made for ongoing technical evaluation and modification as needed to meet emerging management needs. To improve the MRFSS, the committee further recommends that the existing MRFSS program be given a firm deadline linked to sufficient program funding for implementation of this report's recommendations.	All chapters, especially 3-8.	+++++
2. A much greater degree of standardization among state surveys, and between state surveys and the central MRFSS, should be achieved. This will require a much greater degree of cooperation and coordination among the managers of the various surveys.	Chapters 5,6	++++
3. The for-hire sector of marine recreational fisheries should be considered a commercial sector and survey methods and reporting requirements for that sector should therefore be different from those for private anglers.	Chapter 4	+

4.	A comprehensive, universal sampling frame with national coverage should be established. The most effective way to achieve this is through a national registration of all saltwater anglers or through new or existing state saltwater license programs that would allow no exemptions and that provide appropriate contact information from anglers fishing in all marine waters, both state and federal. Any gaps in such a program (for example, a lack of registration in a particular region or mode, exemptions of various classes of anglers, and so on) would compromise the use of the sampling frame and hence the quality of the survey program. An updated, complete registration list would greatly improve sampling efficiency in terms of time and cost. Although these savings might not cover the entire cost of maintaining such a database, the benefit from the increased quantity and quality of the data would be worth the extra cost, especially if there is an associated increase in public confidence in the final estimates.	Chapters 2, 3	+++
5.	Future telephone surveys should be based on the above universal sampling frame.	Chapter 3	+++++
6.	Charter, party, and other for-hire recreational fishing operations should be required to maintain logbooks of fish landed and kept as well as fish caught and released. Providing the information should be mandatory for continued operation in this sector, and all the information should be verifiable and made available to the survey program in a timely manner.	Chapters 4, 5	+
7.	Additional studies are needed to understand the extent to which fish are kept and inspected as well as the extent of catch not available for inspection to improve the accuracy of catch estimates.	Chapter 4	++
8.	Panel surveys, which contact individual anglers repeatedly through time, should be considered in recreational fishing surveys to gather angler trend data and to improve the efficiency of data collection.	Chapter 3	+

TABLE C.1 Continued

2006 Recommendations	Relevant Chapter(s) in Current Report	Ranking
9. The onsite sampling frame for the MRFSS should be redesigned. The estimation procedure depends critically on the assumption that catch rate does not vary according to the nature of the access point. In particular, small or private-access points that most likely are missed might have different catch rates than larger access points, which would lead to bias in the resulting estimators. In addition, the sampling process requires greater quality control (less latitude on the part of the samplers) than it has at present. See the recommendation below for the establishment of an independent research group to investigate matters such as these.	Chapters 2, 4	++++
10. Dual-frame procedures should be used wherever possible to reduce sample bias. For example, if a state has an incomplete list frame based on licenses, the use of a different sampling frame of the state's residents (e.g., random telephone dialing) would reduce the bias. The existence of a universal frame described above would make this approach unnecessary for offsite sampling.	Chapters 2, 3, 5	++++
11. Internet surveys should be considered for their potential use in recreational fishing surveys, especially in panel surveys as a way for anglers to submit information.	Chapters 3, 4	++
STATISTICAL ESTIMATION ISSUES		
12. The statistical properties of various sampling, data-collection, and data-analysis methods should be determined. Assumptions should be examined and verified so that biases can be properly evaluated.	Chapters 2-5	++++
13. A research group of statisticians should design new analyses based on current developments in sampling theory. These examinations should include experimentation, such as specific sampling of activities like nighttime fishing or fishing from private property, whose current under-representation in the MRFSS sampling has the potential to create bias.	Chapters 3, 5	++++

HUMAN DIMENSIONS		
14. An independent national trip and expenditure survey should be developed to support economic valuation studies, impact analyses, and other social and attitudinal studies. The sampling and survey procedures of the independent survey should be designed for the purpose of social and economic, not biological, analyses.	Chapter 6	N/A
15. Add-on surveys for human dimensions should be continued, but in a more focused way than is done currently to target specific management needs and to supplement the national data as needed.	Chapters 3, 4	N/A
16. The national database on marine recreational fishing sites and their characteristics should be enhanced to support social, economic, and other human dimensions analysis. Sites should be defined at levels as fine as possible. The data set should include site characteristics that matter to anglers in making fishing choices, such as boat ramps, facilities, natural amenities, parking, size and type (beach, pier, launch point, and so forth). To account for changes in the number and patterns of trips and the changing characteristics of sites, a periodic updating of the data should be conducted.	Chapters 3, 4	+++++
PROGRAM MANAGEMENT AND SUPPORT		
17. A permanent and independent research group should be established and funded to continuously evaluate the statistical design and adequacy of recreational fishery surveys and to guide necessary modifications or new initiatives. Human dimensions expertise should be included as well.	Chapter 5	++++
18. Additional funding is needed for a survey office devoted to the management and implementation of marine recreational surveys, including coordination between surveys conducted in various state and federal agencies.	Chapters 5, 6	++++

TABLE C.1 Continued

2006 Recommendations	Relevant Chapter(s) in Current Report	Ranking
COMMUNICATION AND OUTREACH		
19. Outreach and communication should be improved in several ways. The MRFSS managers should advise anglers and data users on the constraints that apply to the use of the data for various purposes. Magers and anglers also should be informed clearly about any limitations of the data.	Chapter 7	++
20. Outreach and communication should be institutionalized as part of an ongoing program, so that their importance is acknowledged and appropriate expertise can be developed.	Chapter 7	++
21. Angler associations should be engaged as partners with survey managers through workshops, data collection, survey design, and participation in survey advisory groups. Many NRC and other reports stress the importance of making use of local and traditional knowledge, capacity building, and involving local communities in knowledge-gathering and dissemination activities. Those recommendations apply, as well, to the recreational fishing community.	Chapter 7	+

Appendix D

Excerpt from Magnuson-Stevens Fishery Conservation and Management Reauthorization Act of 2006

TITLE II—INFORMATION AND RESEARCH

SEC. 201. RECREATIONAL FISHERIES INFORMATION.

Section 401 (16 U.S.C. 1881) is amended by striking subsection (g) and inserting the following:

"(g) RECREATIONAL FISHERIES.—

"(1) FEDERAL PROGRAM.—The Secretary shall establish and implement a regionally based registry program for recreational fishermen in each of the 8 fishery management regions. The program, which shall not require a fee before January 1, 2011, shall provide for—

"(A) the registration (including identification and contact information) of individuals who engage in recreational fishing—

"(i) in the Exclusive Economic Zone;

"(ii) for anadromous species; or

"(iii) for Continental Shelf fishery resources beyond the Exclusive Economic Zone; and

"(B) if appropriate, the registration (including the ownership, operator, and identification of the vessel) of vessels used in such fishing.

"(2) STATE PROGRAMS.—The Secretary shall exempt from registration under the program recreational fishermen and charter fishing vessels licensed, permitted, or registered under the laws of a State if the Secretary determines that information from the State program is suitable for the Secretary's use or is used to assist in completing marine recreational fisheries statistical surveys, or evaluating the effects of proposed conservation and management measures for marine recreational fisheries.

"(3) DATA COLLECTION.—

"(A) IMPROVEMENT OF THE MARINE RECREATIONAL FISHERY STATISTICS SURVEY.— Within 24 months after the date of enactment of the Magnuson-Stevens Fishery Conservation and Management Reauthorization Act of 2006, the Secretary, in consultation with representatives of the recreational fishing industry and experts in statistics, technology, and other appropriate fields, shall establish a program to improve the quality and accuracy of information generated by the Marine Recreational Fishery Statistics Survey, with a goal of achieving acceptable accuracy and utility for each individual fishery.

"(B) NRC REPORT RECOMMENDATIONS.—The program shall take into consideration and, to the maximum extent feasible, implement the recommendations of the National Research Council in its report Review of Recreational Fisheries Survey Methods (2006), including—

"(i) redesigning the Survey to improve the effectiveness and appropriateness of sampling and estimation procedures, its applicability to various kinds of management decisions, and its usefulness for social and economic analyses; and

"(ii) providing for ongoing technical evaluation and modification as needed to meet emerging management needs.

"(C) METHODOLOGY.—Unless the Secretary determines that alternate methods will achieve this goal more efficiently and effectively, the program shall, to the extent possible, include—

"(i) an adequate number of dockside interviews to accurately estimate recreational catch and effort;

"(ii) use of surveys that target anglers registered or licensed at the State or Federal level to collect participation and effort data;

"(iii) collection and analysis of vessel trip report data from charter fishing vessels; and

"(iv) development of a weather corrective factor that can be applied to recreational catch and effort estimates.

"(D) DEADLINE.—The Secretary shall complete the program under this paragraph and implement the improved Marine Recreational Fishery Statistics Survey not later than January 1, 2009.

"(4) REPORT.—Within 24 months after establishment of the program, the Secretary shall submit a report to Congress that describes the progress made toward achieving the goals and objectives of the program.."

Appendix E

Survey Instruments

Puerto Rico

2016 ACCESS-POINT ANGLER INTERCEPT SURVEY – Puerto Rico **OMB NO. 0648-0659 (EXP. 02/29/2016)**

2. ASSIGNMENT NO. 1

6. INTERVIEWER TIME *(use 2400 clock)* ☐☐☐☐ *Time this interview was completed*

3. INTERVIEWER ID ☐☐☐☐

7. STATE CODE ☐☐ 8. COUNTY CODE ☐☐☐ 9. SITE CODE ☐☐☐☐

4. YR/MO/DAY 2016 ☐☐/☐☐

10. INTERVIEW STATUS (Key Item = *)
- 1 ☐ **Questionnaire Complete**
- 2 ☐ Refused Non-Key Item
- 5 ☐ Refused Key Item

5. INTERCEPT NO. ☐☐

READ PRIVACY ACT: This study is being conducted in accordance with the privacy act of 1974. You are not required to answer any question that you consider to be an invasion of your privacy.

***11. Would you say you were fishing from...**

SH:
- 0 ☐ Pier
- 1 ☐ Dock
- 2 ☐ Jetty, Breakwater, Breachway
- 3 ☐ Bridge, Causeway
- 4 ☐ Other Man-made Structure (*Specify*)________
- 5 ☐ Beach or Bank (*For Beach Bank only – additional hours required in Q16*)

HB 6 ☐ Head Boat

CH 7 ☐ Charter Boat

PR:
- 8 ☐ Private Boat
- 9 ☐ Rental Boat

***12. Was most of your *(specify mode)* fishing effort today in the ... (Select only one)**
- 1 ☐ Ocean/gulf/open bay
- 2 ☐ Sound (*Other than those specified*)
- 3 ☐ River (*Other than those specified*)
- 4 ☐ Bay (*Other than those specified*)
- 5 ☐ Other (*Specify*) ________
- N ☐ Boqueron Estuary
- O ☐ Guanica Estuary
- P ☐ Guayanilla Estuary
- Q ☐ Humacao Estuary
- R ☐ Jobos Estuary
- S ☐ San Juan Estuary
- T ☐ Tortuguero Estuary

(2 through T: Code Q13 as "8.")

*BOX A. **SHORE mode only** - If Q12 is "ocean/gulf/open bay": code Q13 as "3", 10 miles or less. If Q12 is "2" through "T", code Q13 as "8", Does not apply.*

***13. Was that**
- 3 ☐ 10 Miles or Less from Shore
- 8 ☐ Does not apply
- 4 ☐ More than 10 Miles

14. What type of gear was primarily used? (Select only one)

01 ☐ Hook and Line	07 ☐ Trap
02 ☐ Dip Net, A-frame	08 ☐ Spear
03 ☐ Cast Net	09 ☐ Hand
04 ☐ Gill Net	10 ☐ Other (*Specify*)________
05 ☐ Seine	98 ☐ Unknown
06 ☐ Trawl	99 ☐ Refused

15a. To the nearest half-hour, how many hours have you spent *(specify mode)* fishing today? That is, how many hours have you actually spent with your gear in the water?

☐☐.☐ *Code as "99.9" if DK or Refused*

15b. [If NOT SH, ask] To the nearest half-hour, how many hours have you spent on the boat, away from the dock, today?

☐ Not Applicable – SH mode ("88.8")

☐☐.☐ *Code as "99.9" if DK or Refused*

16. *(Ask if Beach or Bank)* How many additional hours do you expect to fish from shore today? That is, how many *more* will you actually have your gear in the water?

☐☐.☐ Additional BB hours (*only if Q11 =5*)

☐ Not fishing from beach or bank

17. Were you fishing for any particular kinds of fish today? *If Yes, what kinds?*

☐ No Particular Species/ Anything

1st Target________ ☐☐☐☐☐☐

2nd Target________ ☐☐☐☐☐☐

18. Not counting today, within the past 12 months, that is since *(insert month)* of last year, how many days have you gone saltwater sport finfishing in this state or from a boat launched in this state?

☐☐☐ No. of days
- 998 ☐ Don't Know
- 999 ☐ Refused

19. Not counting today, within the past 2 months, how many days?

☐☐ No. of days
- 98 ☐ Don't Know
- 99 ☐ Refused

***20. What is your state and county of residence? *If county unknown ask:* What city or town do you live in?**

☐☐ State Code; Name:________

☐☐☐ County Code; Name:________

21. What is the zip code of your residence?

☐☐☐☐☐ ZIP code
- 99997 ☐ Foreign Country
- 99998 ☐ Don't Know
- 99999 ☐ Refused

22. Do you live in a private residence, or in some type of housing such as a dorm, barracks, nursing home or rooming house?
- 1 ☐ Private Residence
- 2 ☐ Institutional Housing
- 8 ☐ Don't Know
- 9 ☐ Refused

23. At which of the following types of addresses does your household currently receive residential mail? Mark all that apply.

YES	NO	
☐	☐	Street address with a house or building number
☐	☐	Address with a rural route number
☐	☐	U.S. Post Office box (P.O. Box)
☐	☐	Commercial mail box business (such as Mail-boxes, Etc., or Mailboxes Are Us)

☐ Other (Specify) ________

☐ Don't Know ☐ Refused

24. In the event that my supervisor wishes to verify that I have been conducting interviews here today, may I have your name and a phone number?

Angler Name ____________________

D or N PHONE # [ ][ ][ ] – [ ][ ][ ] – [ ][ ][ ][ ]

(If name and/or phone number not given, Q10 = Status 2)

☐ Name and/or phone number not given

☐ Angler aged 16 years or younger (*Check both boxes*)

*25. UNAVAILABLE CATCH Did you catch any fish that are not here for me to look at? For example, any that you may have thrown back or used for bait? NOT GROUP CATCH - Only catch from Angler being interviewed.

Disposition Codes for Q25

1 Thrown back alive	4 Used/plan to use for bait	6 Thrown back dead/plan to throw away
3 Eaten/plan to eat	5 Sold/plan to sell	7 Some other purpose

TYPE 2 RECORDS: *(CATCH UNAVAILABLE IN WHOLE FORM; FILLETS ARE UNAVAILABLE CATCH.)*

Species Name	Species Code	# of Fish	Disp.
1.			
2.			
3.			
4.			
5.			
6.			
7.			

*26. Did you catch any fish while you were fishing that I might be able to look at?

1 ☐ Yes

2 ☐ No - *Code Q27, Q28, Q29 as " Not Applicable"*

3 ☐ Yes, BUT fish on another angler's form – *Fill in interview # where fish are listed*

[ ][ ] *Code Q27, Q28, Q29 as "Not Applicable"*

*27. Did you catch these yourself or did someone else catch some of them?

1 ☐ All Caught by Angler - *Code Q28, Q29 as "Not Applicable"*

2 ☐ *Other Contributors* 8 ☐ Not Applicable

*28. Can you separate out your individual catch?

1 ☐ Yes – *Code 29 as "Not Applicable"*

2 ☐ No 8 ☐ Not Applicable

*29. How many anglers including yourself have their catch here? Please do not include anyone who did not catch fish. Only count those who have their catch here.

[ ][ ] No. of Contributors 88 ☐ Not Applicable

BOX C. If Q11 is SH mode, code Q30 as "888, " and Code Box D as "8."

*30. How many people fished on your boat today?

[ ][ ][ ] No. of People 888 ☐ Shore Mode

BOX D. *If response to Q30 is 1, code as "Not Applicable." Otherwise, is this the first angler from this boat that I have interviewed?

1 ☐ Yes 8 ☐ Not Applicable

2 ☐ No – *Record interview # of 1st angler in the fishing party.* [ ][ ]

*31. AVAILABLE CATCH – ASK: May I look at your fish? What do you plan to do with the MAJORITY of the (species)?

Disposition Codes for Q31

3 Eaten/ plan to eat	5 Sold / Plan to sell	7 Some other purpose
4 Used/plan to use for bait	6 Thrown back dead / Plan to throw away	

TYPE 3 RECORDS: *(INDIVIDUAL CATCH AVAILABLE IN WHOLE FORM)*

Species Name	Species Code	# of Fish	Length (mm)	Weight (kg)	Disp.
1.					
2.					
3.					
4.					
5.					
6.					
7.					
8.					
9.					
10.					
11.					
12.					
13.					
14.					
15.					

North Carolina

2016 APAIS - North Carolina Intercept Form
OMB NO. 0648-0659
EXPIRES 02-29-2016

1. FORM
2. ASSIGNMENT NO. 1
3. INTERVIEWER ID
4. DATE: MM/DD 2016 / /
5. INTERCEPT NO.
6. INTERVIEWER TIME (use 2400 clock) *Time this interview was completed*
7. STATE CODE 37
8. COUNTY CODE
9. SITE CODE
10. INTERVIEW STATUS (Key Item = *)
   - 1 Questionnaire Complete
   - 2 Refused Non-Key Item
   - 5 Refused Key Item

READ PRIVACY ACT: This study is being conducted in accordance with the privacy act of 1974. You are not required to answer any question that you consider to be an invasion of your privacy.

***11. Would you say you were fishing from ...**

SH { 0 Pier; 2 Jetty, Breakwater; 4 Other Man-made Structure (*Specify*) }

1 Dock; 3 Bridge, Causeway; 5 Beach or Bank (*For Beach Bank only - additional hours required in Q16*)

6 Headboat; 7 Charterboat; 8 Private Boat; 9 Rental Boat

11. Did you see any sea turtles while fishing today?
1 Yes, alive; 3 No; 2 Yes, dead

***12. Was most of your *(specify mode)* fishing effort today in the ...**
1 Atlantic Ocean; Other (enter DMF waterbody code and code Q13 as "8").

BOX A. **SHORE mode only** - *If response to Q12 is "Ocean/gulf" code Q13 as "1," 3 miles or less. (If response to Q12 is "2" through "G," code Q13 as "8 Does Not Apply")*

***13. Was that ...**
1 Three Miles or Less From Shore; 8 Waterbody Does Not Apply; 2 More Than Three Miles

13a. Were you fishing near an Artificial Reef?
01 No; 88 SH; If yes, enter reef code 98=unknown

13b. What was the length of the boat used (in feet)? Boat length

14. What type of gear was primarily used? *(Select one only)*
01 Hook and Line; 07 Trap; 02 Dip Net, A-frame; 08 Spear; 03 Cast Net; 09 Hand; 04 Gill Net; 10 Other *(Specify)*; 05 Seine; 98 Unknown; 06 Trawl; 99 Refused

15a. To the nearest half-hour, how many hours have you spent *(specify mode)* fishing today? That is, how many hours have you actually spent with your gear in the water?
No. of Hours — *Codes: If, "Don't Know" = 99.8; If, Refused = 99.9*

15b. [If on boat] To the nearest half-hour, how many hours have you spent on the boat, away from the dock, today?
No. of Hours — *Codes: If, "Don't Know" = 99.8; If, Refused = 99.9*
Not Applicable – SH mode

16. [Ask, only if "Beach" or "Bank"] How many additional hours do you expect to fish from shore today? That is, how many more hours will you actually have your gear in the water?
No. of Hours — *Codes: If, "Don't Know" = 99.8; If, Refused = 99.9*
Not fishing from Beach or Bank

17. What species were you primarily fishing for today?
No Particular Species / Anything
1st Target
2nd Target

18. Not counting today, within the past 12 months, that is since *(insert month)* of last year, how many days have you gone saltwater sport finfishing in this state or from a boat launched in this state?
No. of days; 998 Don't Know; 999 Refused

19. Not counting today, within the past 2 months, how many days?
No. of days; 998 Don't Know; 999 Refused

***20. What is your state and county of residence?** *If county unknown, ask:* **What city or town do you live in?**
State Code; Name:
County Code; Name:

21. What is the ZIP code of your residence?
99997 Foreign Country; 99998 Don't Know; 99999 Refused

22. Do you live in a private residence, or in some type of housing such as a dorm, barracks, nursing home or rooming house?
1 Private Housing; 8 Don't Know; 2 Institutional Housing; 9 Refused

23. At which of the following types of addresses does your household currently receive mail? *(Mark all that apply)*

YES	NO	
		Street address with a house or building number
		Address with a rural route number
		U.S. Post Office Box (P.O. Box)
		Commercial mail box business (e.g., Mailboxes, Etc, UPS Store)

Other *(Specify)*
Don't Know
Refused

23a. Sex
Male
Female

23b. How old were you on your last birthday?
Age
Refused

24. In the event that my supervisor wishes to verify that I have been conducting interviews here today, may I please have your name and phone number?

Angler Name ______________________

D or N
PHONE # [| |] [| |] [| | |]

(If name and/or phone number not given, Q10 = Status 2)
☐ Name and/or phone number not given
☐ Angler aged 16 years or younger (*Check both boxes*)

BOX B. [If headboat ride-along:] Is this one of the anglers you monitored for discard (Type 9) catch? ☐ Yes ☐ No ☐ Not a HB ride

*25. UNAVAILABLE CATCH Did you catch any fish that are not here for me to look at? For example, any that you may have thrown back or used for bait? NOT GROUP CATCH - Only catch from Angler being interviewed.

Disposition Codes for Q25

1 - Thrown back alive	3 - Eaten/plan to eat	5 - Sold/plan to sell	7 - Some other purpose
2 - Thrown back - not legal	4 - Used/plan to use for bait	6 - Thrown back dead/plan to throw away	

TYPE 2 RECORDS: *(CATCH UNAVAILABLE IN WHOLE FORM; FILLETS ARE UNAVAILABLE CATCH.)*

Species Name	Species Code						# of Fish			Disp.
1.										
2.										
3.										
4.										
5.										

*26. Did you catch any fish while you were fishing that I might be able to look at?
1 ☐ Yes
2 ☐ No - *Code Q27, Q28, Q29 as " Not Applicable"*
3 ☐ Yes, BUT fish on another angler's form – *Fill in interview # where fish are listed*
[|] *Code Q27, Q28, Q29 as "Not Applicable"*

*27. Did you catch these yourself or did someone else catch some of them?
1 ☐ All Caught by Angler - *Code Q28, Q29 as "Not Applicable"*
2 ☐ *Other Contributors* 8 ☐ Not Applicable

*28. Can you separate out your individual catch?
1 ☐ Yes – *Code 29 as "Not Applicable"*
2 ☐ No 8 ☐ Not Applicable

*29. How many anglers including yourself have their catch here? Please do not include anyone who did not catch fish. Only count those who have their catch here.
[|] No. of Contributors 88 ☐ Not Applicable

BOX C. If Q11 is SH mode, code Q30 as "888," and Code Box D as "8."

*30. How many people fished on your boat today?
[| |] No. of People 888 ☐ Shore Mode

*BOX D. *If response to Q30 is 1, code as "Not Applicable." Otherwise, is this the first angler from this boat that I have interviewed?*
1 ☐ Yes 8 ☐ Not Applicable
2 [|] No – *Record interview # of 1st angler in the fishing party.*

*BOX E: IS THIS VESSEL ON LIST? YES / NO *NAME OF VESSEL: ______________

☐ Check box if vessel has no name AND record Registration Number to determine "on list" status. If "on list" cannot be confirmed, Q10 = Status 5. *(Note: BOX E must be completed for ALL charter and head boat interviews, regardless of mode of assignment).*

*31. AVAILABLE CATCH – ASK: May I look at your fish? What do you plan to do with the MAJORITY of the (species)?

Disposition Codes for Q31

3 - Eaten/plan to eat	6 - Plan to throw away
4 - Used/plan to use for bait	7 - Some other purpose
5 - Sold/plan to sell	

NOTES/COMMENTS:

TYPE 3 RECORDS: *(INDIVIDUAL CATCH AVAILABLE IN WHOLE FORM)*

Species Name	Species Code						# of Fish			Length (mm)				Weight (kg)						Disp.
1.																	.			
2.																	.			
3.																	.			
4.																	.			
5.																	.			
6.																	.			
7.																	.			
8.																	.			
9.																	.			
10.																	.			
11.																	.			
12.																	.			
13.																	.			
14.																	.			
15.																	.			

Gulf of Mexico

2016 MRIP INTERCEPT SURVEY **OMB NO. 0648-0659 (EXP. 02/29/16)**

2. ASSIGNMENT NO. ☐ Write 1 or 2

6. INTERVIEW TIME (use 2400 clock) ☐☐☐☐ Time the interview was completed

3. INTERVIEWER ID. ☐☐☐☐

4. YR/MO/DAY ☐☐ ☐☐ ☐☐

7. STATE CODE ☐☐ 8. COUNTY CODE ☐☐☐ 9. SITE CODE ☐☐☐☐

5. INTERCEPT NO. ☐☐☐

10. INTERVIEW STATUS (Key Item = *)
1 ☐ Questionnaire Complete
2 ☐ Refused Non-Key Item
5 ☐ Refused Key Item

READ PRIVACY ACT. This study is being conducted in accordance with the privacy act of 1974. You are not required to answer any question that you consider to be an invasion of your privacy.

***11. Would you say you were fishing from........**

1 ☐ Pier, Dock
2 ☐ Jetty, Breakwater, Breachway
SH 3 ☐ Bridge, Causeway
4 ☐ Other Man-made Structure (Specify)________________
5 ☐ Beach or Bank
PC 6 ☐ Headboat — REFER TO Q. 11a
PC 7 ☐ Charterboat — REFER TO Q. 11a
PR 8 ☐ Private or Rental Boat

***11a. Is vessel on Good List** 1 ☐ YES 2 ☐ NO Vessel ID ☐☐☐☐☐☐☐

***12. Was most of your (specify mode) fishing effort today in the......**

1 ☐ Ocean/gulf/open bay
2 ☐ A sound *(Other than those specified)*
3 ☐ River *(Other than those specified)*
4 ☐ Bay *(Other than those specified)*
5 ☐ Other (*Specify*)________________
G ☐ Abermarle/Pamilco Estuary
H ☐ Biscayne Estuary
I ☐ Whitewater Estuary
J ☐ Sarasota Estuary
K ☐ Tampa Estuary
L ☐ Mobile Estuary
M ☐ Atchafalaya Estuary

(2–M: *Code q.13='8'*)

BOX A. Refer to q. 12. If response is NOT "1" code q. 13 as "8", Not applicable

13. Was that** ***West Florida Only

1 ☐ 3 Miles or less from shore 3 ☐ 10 Miles or less from shore
2 ☐ More than 3 Miles 4 ☐ More than 10 miles 8 ☐ Not Applicable

14. What type of gear was primarily used?

01 ☐ Hook and Line 07 ☐ Trap
02 ☐ Dip Net, A-frame 08 ☐ Spear
03 ☐ Cast Net 09 ☐ Hand
04 ☐ Gill Net 10 ☐ Other (Specify)________________
05 ☐ Seine 98 ☐ Unknown
06 ☐ Trawl 99 ☐ Refused

15. To the nearest half-hour, how many hours have you spent (specify mode) fishing today? That is, how many hours have you actually spent with your gear in the water?

☐☐ . ☐ No. of hours

16. How many additional hours do you expect to fish from shore today? That is, how many more hours will you actually have your gear in the water?

☐☐ . ☐ Additional BB Hours

17. Were you fishing for any particular kinds of fish today? ***If Yes,*** **what kinds?**

☐ No Particular Species/Anything

1st Target ________________ ☐☐☐☐☐☐

2nd Target ________________ ☐☐☐☐☐☐

18. Not counting today, within the past 12 months, that is since (insert month) of last year, how many days have you gone saltwater sport finfishing in this state or from a boat launched in this state?

☐☐☐ No. of Days
998 ☐ Don't know
999 ☐ Refused

19. Not counting today, within the past 2 months, how many days?

☐☐ No. of Days
98 ☐ Don't Know
99 ☐ Refused

20. What is your state and county of residence?** ***If county unknown ask: **What city or town do you live in?**

☐☐ State Code; Name ________________
☐☐☐ County Code; Name ________________

21. What is the zip code of your residence?

☐☐☐☐☐ Zip Code
99997 ☐ Foreign Country (Skip to Q 25)
99998 ☐ Don't Know
99999 ☐ Refused

22. Do you live in a private residence, or in some type of housing such as a dorm, barracks, nursing home or rooming house?

1 ☐ Private Residence 8 ☐ Don't know
2 ☐ Institutional Housing 9 ☐ Refused

23. At which of the following types of addresses does your household currently receive residential mail? Mark all that apply.

YES	NO	
☐	☐	Street address with a house or building number
☐	☐	Address with a rural route number
☐	☐	U.S. Post Office box (P.O. Box)
☐	☐	Commerical mail box business (such as Mailboxes etc or Mailboxes Are Us)

☐ Other (Specify) ________________
☐ Don't Know ☐ Refused

24. In the event that my supervisor wishes to verify that I have been conducting interviews here today, may I have your name and a phone number?

ANGLER'S NAME

Name and Phone Number Not Given

D OR N

PHONE # ( ) - -

***25. UNAVAILABLE CATCH. Did you land any fish that are not here for me to look at? For example, any that you may have thrown back or used for bait? IF YES, COMPLETE TYPE 2 RECORD FOR THIS INDIVIDUAL ANGLER, NOT GROUP CATCH. NOTE: FILLETS ARE UNAVAILABLE.**

DISPOSITION CODES FOR Q. 25

1 Thrown back alive/legal	3 Eaten/plan to eat	5 Sold/plan to sell	7 Other purpose
2 Thrown back alive/not legal/legality refused	4 Used for bait/plan to use for bait	6 Thrown back dead/plan to throw away	

TYPE 2 RECORDS: *(INDIVIDUAL CATCH UNAVAILABLE IN WHOLE FORM)*

	SPECIES CODE	# OF FISH	DISP.
1.			
2.			
3.			
4.			
5.			

***26. Did you catch any fish while you were fishing that I might be able to look at?**

1 **Yes**

2 **No** Code q. 27, 28, 29 as "8's,", Not applicable

3 **Yes, BUT fish are on another angler's form** - *Fill in interview # where fish are listed **

Code q. 27, 28, 29 as "8's,", Not applicable

***27. Did you catch these yourself or did someone else catch some of them?**

1 **All caught by Angler** - Code q. 28, 29, as "8's," Not applicable

2 **Other Contributors** 8 **Not applicable**

***28. Can you separate out your individual catch?**

1 **Yes** - Code q. 29 as "88"

2 **No** 8 **Not applicable**

***29. How many anglers including yourself have their catch here? Please do not include anyone who did not catch fish. Only count those anglers who have their catch here.**

No. of Contributors 88 **Not applicable**

***30. How many people fished on your boat today?**

No. of People

888 **Shore Mode**

***BOX D. If response to q. 30 is 1 angler or "88" Shore Mode, check not applicable and skip to q. 31. Otherwise, is this the first angler from this boat that I have interviewed?**

1 **Yes** 8 **Not applicable**

2 **No** - Record interview # of 1st angler in the fishing party

***31. AVAILABLE CATCH. COMPLETE TYPE 3 RECORD BY ASKING: May I look at your fish? What do you plan to do with the MAJORITY of the (species)?**

DISPOSITION CODES FOR Q. 31

3 Eaten/plan to eat	5 Sold/plan to sell	7 Some other purpose
4 Used for bait/plan to use for bait	6 Thrown back dead/plan to throw away	9 Refused

	SPECIES CODE	# OF FISH	LENGTH (mm)	WEIGHT (kg)	DIS
1.				.	
2.				.	
3.				.	
4.				.	
5.				.	
6.				.	
7.				.	
8.				.	
9.				.	
10.				.	
11.				.	
12.				.	
13.				.	
14.				.	
15.				.	

Atlantic States

2016 APAIS - Atlantic Intercept Form
OMB NO. 0648-0659
EXPIRES 02-29-2016

1. FORM

2. ASSIGNMENT NO. 1

3. INTERVIEWER ID

4. DATE: MM/DD 2016 / /

5. INTERCEPT NO.

6. INTERVIEWER TIME (use 2400 clock) *Time this interview was completed*

7. STATE CODE

8. COUNTY CODE

9. SITE CODE

10. INTERVIEW STATUS (Key Item = *)
- 1 Questionnaire Complete
- 2 Refused Non-Key Item
- 5 Refused Key Item

READ PRIVACY ACT: This study is being conducted in accordance with the privacy act of 1974. You are not required to answer any question that you consider to be an invasion of your privacy.

***11. Would you say you were fishing from ...**

SH:
- 0 Pier
- 1 Dock
- 2 Jetty, Breakwater
- 3 Bridge, Causeway
- 4 Other Man-made Structure (*Specify*)
- 5 Beach or Bank (*For Beach Bank only - additional hours required in Q16*)

- 6 Headboat
- 7 Charterboat
- 8 Private Boat
- 9 Rental Boat

***12. Was most of your *(specify mode)* fishing effort today in the ...** *(Select only one)*
- 1 Ocean/gulf

Q13 as 8:
- 2 Sound (*other than listed*)
- 3 River (*other than listed*)
- 4 Bay (*other than listed*)
- 5 Other (*Specify*)
- V Cape Cod Bay
- A Narragansett Estuary
- B Buzzard's Bay Estuary
- C Long Island Estuary
- D Hudson/Raritan Estuary
- E Delaware Estuary
- F Chesapeake Estuary
- G Albemarle/Pamlico Estuary

BOX A. **SHORE mode only** - *If response to Q12 is "Ocean/gulf" code Q13 as "1," 3 miles or less. (If response to Q12 is "2" through "G," code Q13 as "8 Does Not Apply")*

***13. Was that ...**
- 1 Three Miles or Less From Shore
- 2 More Than Three Miles
- 8 Waterbody Does Not Apply

14. What type of gear was primarily used? *(Select one only)*
- 01 Hook and Line
- 02 Dip Net, A-frame
- 03 Cast Net
- 04 Gill Net
- 05 Seine
- 06 Trawl
- 07 Trap
- 08 Spear
- 09 Hand
- 10 Other (*Specify*)
- 98 Unknown
- 99 Refused

15a. To the nearest half-hour, how many hours have you spent *(specify mode)* fishing today? That is, how many hours have you actually spent with your gear in the water?

No. of Hours

Codes: If, "Don't Know" = 99.8; If, Refused = 99.9

15b. [If on boat] To the nearest half-hour, how many hours have you spent on the boat, away from the dock, today?

No. of Hours

Codes: If, "Don't Know" = 99.8; If, Refused = 99.9

Not Applicable – SH mode

16. [Ask, only if "Beach" or "Bank"] How many additional hours do you expect to fish from shore today? That is, how many more hours will you actually have your gear in the water?

No. of Hours

Codes: If, "Don't Know" = 99.8; If, Refused = 99.9

Not fishing from Beach or Bank

17. What species were you primarily fishing for today?

No Particular Species / Anything

1st Target

2nd Target

18. Not counting today, within the past 12 months, that is since *(insert month)* of last year, how many days have you gone saltwater sport finfishing in this state or from a boat launched in this state?

No. of days
- 998 Don't Know
- 999 Refused

19. Not counting today, within the past 2 months, how many days?

No. of days
- 998 Don't Know
- 999 Refused

***20. What is your state and county of residence?** *If county unknown, ask:* **What city or town do you live in?**

State Code; Name:

County Code; Name:

21. What is the ZIP code of your residence?
- 99997 Foreign Country
- 99998 Don't Know
- 99999 Refused

22. Do you live in a private residence, or in some type of housing such as a dorm, barracks, nursing home or rooming house?
- 1 Private Housing
- 2 Institutional Housing
- 8 Don't Know
- 9 Refused

23. At which of the following types of addresses does your household currently receive mail? *(Mark all that apply)*

YES	NO	
		Street address with a house or building number
		Address with a rural route number
		U.S. Post Office Box (P.O. Box)
		Commercial mail box business (e.g., Mailboxes, Etc., UPS Store)

Other (*Specify*)

Don't Know

Refused

24. In the event that my supervisor wishes to verify that I have been conducting interviews here today, may I please have your name and phone number?

Angler Name ______________________

D or N
PHONE # ___ ___ ___ | ___ ___ ___ | ___ ___ ___ ___

(If name and/or phone number not given, Q10 = Status 2)
☐ Name and/or phone number not given
☐ Angler aged 16 years or younger (*Check both boxes*)

BOX B. [If headboat ride-along:] Is this one of the anglers you monitored for discard (Type 9) catch? ☐ Yes ☐ No ☐ Not a HB ride

***25. UNAVAILABLE CATCH Did you catch any fish that are not here for me to look at? For example, any that you may have thrown back or used for bait? NOT GROUP CATCH - Only catch from Angler being interviewed.**

Disposition Codes for Q25

1 - Thrown back alive	3 - Eaten/plan to eat	5 - Sold/plan to sell	7 - Some other purpose
2 - Thrown back - not legal	4 - Used/plan to use for bait	6 - Thrown back dead/plan to throw away	

TYPE 2 RECORDS: ***(CATCH UNAVAILABLE IN WHOLE FORM; FILLETS ARE UNAVAILABLE CATCH.)***

Species Name	Species Code	# of Fish	Disp.
1.			
2.			
3.			
4.			
5.			

***26. Did you catch any fish while you were fishing that I might be able to look at?**

1 ☐ Yes
2 ☐ No - *Code Q27, Q28, Q29 as " Not Applicable"*
3 ☐ Yes, BUT fish on another angler's form – *Fill in interview # where fish are listed*
☐☐ *Code Q27, Q28, Q29 as "Not Applicable"*

***27. Did you catch these yourself or did someone else catch some of them?**

1 ☐ All Caught by Angler - *Code Q28, Q29 as "Not Applicable"*
2 ☐ *Other Contributors* 8 ☐ Not Applicable

***28. Can you separate out your individual catch?**

1 ☐ Yes – *Code 29 as "Not Applicable"*
2 ☐ No 8 ☐ Not Applicable

***29. How many anglers including yourself have their catch here? Please do not include anyone who did not catch fish. Only count those who have their catch here.**

☐☐ No. of Contributors 88 ☐ Not Applicable

BOX C. If Q11 is SH mode, code Q30 as "888," and Code Box D as "8."

***30. How many people fished on your boat today?**

☐☐☐ No. of People 888 ☐ Shore Mode

BOX D. *If response to Q30 is 1, code as "Not Applicable." Otherwise, is this the first angler from this boat that I have interviewed?

1 ☐ Yes 8 ☐ Not Applicable
2 ☐☐ No – *Record interview # of 1st angler in the fishing party.*

***BOX E: IS THIS VESSEL ON LIST? YES / NO *NAME OF VESSEL:** ______________________

☐ **Check box if vessel has no name AND record Registration Number to determine "on list" status. If "on list" cannot be confirmed, Q10 = Status 5. *(Note: BOX E must be completed for ALL charter and head boat interviews, regardless of mode of assignment).***

***31. AVAILABLE CATCH – ASK: May I look at your fish? What do you plan to do with the MAJORITY of the (species)?**

Disposition Codes for Q31

3 - Eaten/plan to eat 6 - Plan to throw away
4 - Used/plan to use for bait 7 - Some other purpose
5 - Sold/plan to sell

NOTES/COMMENTS:

TYPE 3 RECORDS: ***(INDIVIDUAL CATCH AVAILABLE IN WHOLE FORM)***

Species Name	Species Code	# of Fish	Length (mm)	Weight (kg)	Disp.
1.				.	
2.				.	
3.				.	
4.				.	
5.				.	
6.				.	
7.				.	
8.				.	
9.				.	
10.				.	
11.				.	
12.				.	
13.				.	
14.				.	
15.				.	

Appendix F

2014 Calibration Workshops

Appendix 1. Detailed Implementation Steps for the Calibration Methods Proposed During the Workshop.

Summary Report: NOAA Calibration Methods Workshop - Charleston, SC
September 8-10, 2014
Lynne Stokes, Ken Pollock, Ginny Lesser
December 18, 2014

The new MRIP Access point survey has replaced the original MRFSS Access Point Survey. A variety of design changes have been made. One major consequence is that the new survey covers the fishing day more effectively than the original MRFSS Access Point Survey. Because the time series of recreational catch rate estimates form the basis of so many important fisheries stock assessments, there is the need to develop methods which "calibrate" the original time series of MRFSS estimates to the new time series of MRIP estimates. This is a difficult statistical estimation and prediction issue because both surveys were not run in parallel in any years (except for one pilot test in NC). The new estimates can be very different from the old estimates causing an abrupt change in the time series.

The purpose of this document is to outline the steps involved in implementing several model dependent calibration approaches to re-estimate catch that were discussed at the Charleston workshop. In addition, we discuss their assumptions. The first two methods use ideas of ratio estimation and assume that the major changes between the two surveys are due to a better temporal coverage of the fishing day in the new MRIP survey. The third method is a regression prediction modeling approach that will take longer to develop. None of these methods incorporate any analysis of spatial patterns or include time series methods, which might improve estimates. This would be worth exploring to determine if time series or small area estimation techniques for this short time series might provide improved estimates.

1. Direct Catch Ratio Adjustment
    - Steps in approach (for each subregion, state, mode, species.):
        i. Define peak period for each of the domains (excluding species). Peak period is defined using two criteria: 1) the contiguous range of hours during which weighted hourly proportions of total trips in the MRFSS years (prior to 2013) were greater than or equal to the corresponding weighted hourly proportions of total trips in 2013, and 2) the peak period accounted for at least 75% of the intercept data (trips) in the MRFSS years.
        ii. Estimate peak and total catch using the 2013 data based on the MRIP survey method where both the peak and total fishing periods were sampled adequately. Denote these by $c_{p,2013}$ and $c_{total,2013}$, respectively.

iii. Calculate the ratio $R_{2013} = c_{total,w2013}/c_{p,2013}$. This estimate and its large sample variance, based on standard Taylor series methods, can be calculated from survey sampling software packages such as SAS.

iv. Denote the estimator of catch based on the MRFSS method during the peak period in earlier year y (e.g., y = 2012, 2011, etc.) by $c_{p,y}$. Then the estimator of adjusted total catch for year y (i.e., a prediction of what would have been obtained if MRIP had been run) will be calculated as the product of the ratio from year 2013 and the catch for the peak period in year y; i.e.,

$$c_{tot,y} = R_{2013} * c_{p,y}.$$

iv. The variance of the adjusted catch $c_{tot,y}$ can be calculated using the expression for the variance of a product of two independent random variables introduced by Goodman (1960): .

$$var(c_{tot,y}) = var(R_{2013})(c_{p,y})^2 + var(c_{p,y})(R_{2013})^2 - var(R_{2013})var(c_{p,y})$$

By substituting estimates for each of the components in this equation, the variance can be estimated.

- Assumptions:
  - i. Relative distribution of catch throughout day (i.e., between peak and total) is constant between 2013 and the year that is being adjusted for each domain

- Advantages:
  - i. Simple to apply.

- Disadvantages:
  - i. Information that is available for non-peak hours are not used.

- Two variations of this approach:
  - i. Keep a fixed peak time the same (note this will vary by state and mode)
  - ii. Use different peak times (allow this to vary by state, mode and year since this was allowed to vary in these groups)

2. Complex Ratio Method Based on Fishing Effort Distributions

- Steps in approach (for each subregion, state, mode, species etc.):
  - i. The 2013 daily relative distribution of total fishing effort is obtained and also the relative distribution of total fishing effort data for the year to be compared to (for example, for y = 2012, 2011, etc.). Total fishing effort is estimated as the fishing effort estimate from separate telephone surveys (CHTS, FHS) that is subsequently expanded by coverage correction factors estimated from APAIS.

ii. The 2013 sampling weights are then adjusted (up or down weighted) so that the 2013 relative distribution matches the year y relative distribution. This is to be done by using discrete temporal bins with the exact bin widths yet to be determined. The adjustments made to the 2013 sample weights are a ratio style adjustment of the form:

$$w_{dti|y} = w_{dti} * \frac{\hat{p}_{dt,y}}{\hat{p}_{dt,2013}}$$

where w_{dti} is the unadjusted 2013 sample weight for angler-trip *i* in time bin *t* in subregion, state, mode domain *d*,
$\hat{p}_{dt,2013}$ is the original 2013 weighted proportion for time bin *t* of total trips in domain *d*,
$\hat{p}_{dt,y}$ is the year *y* weighted proportion for time bin *t* of total trips in domain *d*, and
$w_{dti|y}$ is the 2013 sample weight for angler-trip *i* in time bin *t* in domain *d* adjusted to year *y*.

From initial evaluations of bin width, it appears that a 3-hour bin is the smallest bin that results in no data gaps or mismatches in 2013 (data present in a bin in a prior year but not in 2013) for all state by mode domains. However, additional work could be done to fine tune bin widths for each domain cell.

iii. Use the MRIP survey method to estimate catch for the complete 2013 data and denote it by c_{2013}. Also calculate catch for the 2013 data weighted to match the truncated distribution of effort for year y data (step ii above), and denote this estimator by $c_{tr,2013}$
iv. Calculate the ratio of 2013 complete to truncated catch based on the MRIP survey; i.e., $R_{c/tr,2013} = c_{2013}/c_{tr,2013}$.
v. Multiply this ratio by the year y estimate of catch c_y to obtain the adjusted year y catch estimate (i.e. what would have been obtained if MRIP survey had been run) $c_{y,adj} = R_{c/tr,2013} * c_y$.
vi. A similar approach can be used to adjust all other years one by one or alternately down weight 2013 compared to the pooled temporal distribution of all other years and get one overall ratio which can be used to adjust all the years.
vii. Explore computation of the variances of the calibrated estimates by either using a bootstrap or delta method.

- Assumptions:
  i. Assumptions for this approach, such as constant relative distribution of trip/catch characteristics between years in the comparison/adjustment, must be investigated to determine if assumptions are met and will lead to consistent estimators.

- Advantages:
    i. Information that is available for non-peak hours are used unlike in the previous method.

- Disadvantages:
    i. Information from non-peak hours will be limited and may be highly variable or impacted by incomplete coverage compared to information from peak hours.
    ii. The assumptions under which this estimator will be consistent (that is, will provide an unbiased estimate for a sufficiently large sample size) are unknown at this time. For example, if the (strong) assumption needed for Method 1 is assumed, the estimator will still not necessarily be consistent.

- Other ideas to consider as variations of above
    i. Recalculate catch after effort has been readjusted. Therefore, both catch and effort are readjusted. The calibration methods make use of the MRIP public-use or micro datasets. The records included in these datasets come from APAIS. However, the sample weights in these datasets include a post-stratification adjustment such that the sum of the sample weights equals the MRIP estimate of total effort in domain cells defined by year, subregion, state, wave, mode, and area. To more fully approximate the effect of temporal coverage changes on catch, the MRIP estimates of total effort must be recalculated since they also include coverage correction factors estimated from APAIS. Once total effort has been recalculated, sample weights may be post-stratified to the new effort totals, and then revised catch estimates may be calculated as weighted sums using sample weights that have been adjusted to both a prior year daily distribution of effort as well as the resultant new effort total.
    ii. Apply temporal distribution either year-by-year or as an average across a range of years (say 2004-2012). Then multiply this ratio by MRFSS estimates of catch in previous years. NOTE: If use each year separately, then there is no assumption that the relative distribution of catch is constant throughout the day across years, only the two years that are compared. So if only one year violates this assumption, then conducting an aggregate analysis could bias the estimator for the other years, while if it was done separately, only it would be biased by that assumption violation. Conversely, using a multi-year average distribution may work to smooth results in cases where annual level distributions may be more variable.

3. Regression Model-Based Approach
    - Steps in approach:
        i. Develop a regression model using 2013 intercept data (perhaps other years as well) to predict and classify trips into either morning, peak, or evening as predicted from

their characteristics, such as type of catch and other demographic and behavior characteristics of the anglers that are available from the intercept questionnaire. Cross-validation could be used to check the model. For example, one could use approximately 75% of the data to develop the model. Then Bayes' Information Criterion (or other model fit statistic) could be used to develop the best fitting model. Once the model is built, the remaining 25% of the data could be used to predict the response variable. A statistic, such as the Press statistic, could be calculated to document how well the model is predicting the response categories. A replication approach might also be considered to look at model robustness or stability.

ii. Use the model to predict Morning, Peak and Evening trips for 2012, 2011, etc. These classifications won't be "true" morning, peak, and evening categories, since they won't be aiming to identify when the trip took place. Rather, they will be trying to predict when a trip is similar, based on catch and demographic and behavior characteristics of anglers, to trips in 2013 in those categories.

iii. Determine the proportion of Morning, Peak, and Evening trips in 2013. Adjust the 2012, 2011, data so that the Morning, Peak, Evening proportions are identical to the 2013 data. These are adjusted proportions. In addition to 2013 data, control proportions for prior years may be developed using trip time data from the CHTS and FHS effort surveys, which would be available for a range of years prior to 2013.

iv. This new weight, the inverse of the 'adjusted proportions', is multiplied by the existing weights for 2012, 2011, etc. to create the adjusted weight.

v. Data are now analyzed using the adjusted weights.

vi. A bootstrap method could be used to calculate variances.

- Assumptions:
  i. Reasonable predictive model can be developed using 2013 data to reasonably predict catch period type (i.e., Morning, Peak, and Evening).
  ii. The demographic characteristics of the angler/catch predict the characteristics of the catch through a "label" we are assigning about time of day.
  iii. Assumes that true time and latent time are identical in 2013 (see below for definition of latent.)

- Disadvantages:
  i. More work is required to develop the prediction model.

  The model is not designed to predict the observable characteristic (time of day), but is rather predicting whether the trip "resembles" a trip made during that time of day, which is a latent variable. Because of this, the model checking done on the 2013 data to see how well the model works is not like the target years, since we can't observe the latent variable even for 2013. It may be that some of the trips made in the morning in 2013 do not resemble morning trips, and yet the model will be examined for its accuracy in predicting true time. If we were really interested in predicting true time, we would simply use the true time as a predictor in previous years!

- Advantages
    i. A number of important explanatory variables can be incorporated in the model to better predict trips.
    ii. Approach incorporates the calibration into the sample weights, which maintains the current usability of MRIP public-use datasets for analysts.

- Other comments:
    i. As more data is collected using the MRIP design, the model development should be repeated to improve prediction.

Catch can also be added to model, but need to be careful of applying 2013 year affects to previous years.

References:

Goodman, Leo A., "On the exact variance of products," *Journal of the American Statistical Association*, December 1960, 708–713.

Appendix 2. Recommended Interim Calibration Approach, suggested for use in Assessments Conducted in Winter 2014/15.

October 30, 2014
Summary Report: Recommended NOAA Calibration Method
Lynne Stokes, Ken Pollock, Ginny Lesser

Introduction

The new MRIP Access Point Angler Intercept Survey (APAIS) has replaced the original MRFSS Access Point Survey. A variety of design changes have been made. One major consequence is that the new survey covers the fishing day more effectively than the original MRFSS Access Point Survey. Because the time series of recreational catch rate estimates form the basis of so many important fisheries stock assessments, there is the need to develop methods which "calibrate" the original time series of MRFSS estimates to the new time series of MRIP estimates. This is a difficult statistical estimation and prediction issue because the two surveys were not run in parallel in any years (except for one pilot test in NC). The new estimates can be very different from the old estimates causing an abrupt change in the time series. Three methods of producing a calibration were suggested at the workshop in Charleston, SC held in September. Since that time, the statistical consultants have worked on investigating the properties of the three methods, and John Foster has implemented two of the three methods for some areas/species, in order to see how they perform. The purpose of this document is to describe our recommended method and to explain our choice.

Our recommendation

Our recommendation at this time is to use the method that was referred to as "Method 1" at the workshop. Our decision is based on two main factors. One is that the method is the easiest to explain and to understand of the three methods. It is based on an assumption that the ratio of catch in the peak period to total catch is stable over time. The method referred to as "Method 2" at the workshop is also a ratio method, but it is more complex (a negative feature) and uses the data from prior years more fully (a positive feature). Our reluctance to recommend Method 2 at this time is that we have not yet been able to determine the assumptions under which this estimator is consistent. For example, the strong assumptions required for consistency of the method 1 estimator are not sufficient to ensure consistency of the method 2 estimator. It is also clear that the method 2 estimator requires estimation of more parameters than Method 1. As a result, we are not confident that the one year of new MRIP APAIS estimates available at this time will be sufficient. Finally, Method 3 considered at the conference is a regression prediction modeling approach that will take longer to develop and also need more data. (It is the one method not yet applied to any of the data by John Foster.)

Description of the method

Here we describe the basic assumption used to justify Method 1, and then outline the steps required for implementation. First, the justification of the method requires the assumption that in years previous to 2013, there is a period of the day that can be considered to have been fully covered by the MRFSS survey, and that the bias in its estimates occurs due to undercoverage in the non-peak periods. This is a very strong, but necessary assumption for this method. Second, the method requires the assumption that the ratio of peak catch to total catch stays constant across years for subregion, state, mode, and species. So for each of these domains, the calibrated total catch for year *y* is made as

$$\hat{C}_{tot,y} = \hat{R}_{2013}\hat{C}_{p,y} \quad (1)$$

where $\hat{C}_{p,y}$ is the estimated peak-period catch for year *y* calculated from reweighted MRFSS data and $\hat{R}_{2013} = \hat{C}_{tot,2013} / \hat{C}_{p,2013}$ is the ratio of the total to peak catch for year 2013, which is calculated from MRIP data. $\hat{C}_{tot,y}$ is thus our estimate of the catch total for the domain that would have been estimated if MRIP had been conducted in year *y*.

The steps in producing this estimate are outlined below.

Step 1. Define peak period for each of the domains (subregion, state, mode). In the pilot implementation by John Foster, peak period was defined using two criteria: 1) the contiguous range of hours during which weighted hourly proportions of total trips in the MRFSS years (prior to 2013) were greater than or equal to the corresponding weighted hourly proportions of total trips in 2013, and 2) the peak period accounted for at least 75% of the intercept data (trips) in the MRFSS years.

Step 2. Calculate $\hat{C}_{p,y}$, the catch in the peak period for all years $y < 2013$ for which calibration is needed.

Step 3. Estimate peak and total catch using the 2013 data based on the MRIP survey method where both the peak and total fishing periods were sampled adequately. Calculate its ratio $\hat{R}_{2013}$.

Step 4. Calculate the estimator $\hat{C}_{tot,y}$ shown in (1).

The variance of this estimator can be calculated using standard statistical methods.

Discussion

There are at least three substantial criticisms possible for this method. First is that the method uses none of the data collected outside the peak period in years prior to 2013. The second is that the method requires an assumption that the ratio of catch in the peak period to total catch is constant across years. We are not sure if this is defensible from a scientific point of view. Third, the method assumes that the estimate of total catch for the peak period made from the reweighted MRFSS data in years prior to 2013 is unbiased. On the other hand, some type of unverifiable assumption will be necessary in order to carry out any calibration because of the lack of side-by-side data collection for the MRIP and MRFSS APAIS sampling designs.

Some variations on Method 1 are possible. For example, the choice of how the peak period is defined will affect the estimates. Peak can be determined individually for each year or based on an aggregation of years and/or domains. We believe that this definition will be difficult to specify in advance, and must be based on characteristics of the data.

We recommend that investigation continue on the remaining two methods. It is possible that one of them will be determined to be better at some future date.

Appendix G

Acronym List

AAPOR	American Association of Public Opinion Research
ABC	Acceptable Biological Catch
ABS	Address-based sampling
ACCSP	Atlantic Coast Cooperative Statistics Program
ACL	Annual Catch Limit
ACS	American Community Survey
ACT	Annual Catch Target
ADF&G	Alaska Department of Fish and Game
ALDS	Angler License Directory Survey
APAIS	Access Point Angler Intercept Survey
ASMFC	Atlantic States Marine Fishery Commission
CDS	Computerized Delivery Sequence
CET	Communications and Education Team
CH	Charter fishing boat mode
CHTS	Coastal Household Telephone Survey
CPUE	Catch per Unit Effort
DMR	Discard Mortality Rate
ER	Electronic Reporting
ESC	Executive Steering Committee
FCC	Federal Communications Commission
FCMA	(Magnuson-Stevens) Fishery Conservation and Management Act

FES	Fishing Effort Survey
FHS	For-Hire Survey
FIN	Fishery Information Network
FWS	(U.S.) Fish and Wildlife Service
GSMFC	Gulf States Marine Fishery Commission
HB	Head boat (party boat) mode
HMS	Highly Migratory Species
IPHC	International Pacific Halibut Commission
LPBS	Large Pelagics Biological Survey
LPS	Large Pelagics Survey
MRFSS	Marine Recreational Fisheries Statistics Survey
MRIP	Marine Recreational Information Program
MSFCMA	Magnuson-Stevens Fishery Conservation and Management Act (sometimes FCMA)
NAS	National Academy of Sciences
NGO	nongovernmental organization
NMFS	National Marine Fisheries Service
NOAA	National Oceanic and Atmospheric Administration
NPFMC	North Pacific Fishery Management Council
NRC	National Research Council
NSAR	National Saltwater Angler Registry
ODFW	Oregon Department of Fish and Wildlife
OFL	overfishing limit
ORBS	Oregon Recreational Boat Survey (of ODFW)
OSP	Ocean Sampling Program (of ODFW)
PPS	probability proportional to size
PR	Private/rental boat mode
PSE	Proportional Standard Error (reported in percent)
PSMFC	Pacific States Marine Fishery Commission
PSSP	Puget Sound Sampling Program (of WDFW)
PSU	primary sampling unit
QA	quality assurance
QC	quality control

RARA	Report of Assessment and Research Activity
RDD	random digit dialing
RecFIN	Recreational Fishery Information Network
SAFE	Stock Assessment and Fishery Evaluation
SH	Shore mode
SOW	statement of work
SR	Site Registry
SSC	Scientific and Statistical Committee (of regional fishery management councils)
SSU	secondary sampling unit
TPDW	Texas Parks and Wildlife Department
TSU	tertiary sampling unit
USFWS	U.S. Fish and Wildlife Service (sometimes FWS)
WPacFIN	Western Pacific Fishery Information Network